Hydrogeology: A Global Scenario

Hydrogeology: A Global Scenario

Edited by **William Sobol**

New York

Published by Callisto Reference,
106 Park Avenue, Suite 200,
New York, NY 10016, USA
www.callistoreference.com

Hydrogeology: A Global Scenario
Edited by William Sobol

International Standard Book Number: 978-1-63239-425-5 (Hardback)

Contents

Preface VII

Chapter 1 **Hydrogeology of Karstic Area** 1
Haji Karimi

Chapter 2 **Hydrogeological Significance of Secondary Terrestrial
Carbonate Deposition in Karst Environments** 43
V.J. Banks and P.F. Jones

Chapter 3 **Significance of Hydrogeochemical Analysis in
the Management of Groundwater Resources:
A Case Study in Northeastern Iran** 79
Gholam A. Kazemi and Azam Mohammadi

Chapter 4 **A Review of Approaches for Measuring Soil
Hydraulic Properties and Assessing the Impacts
of Spatial Dependence on the Results** 97
Vincenzo Comegna, Antonio Coppola,
Angelo Basile and Alessandro Comegna

Chapter 5 **Hydrogeological-Geochemical Characteristics of
Groundwater in East Banat, Pannonian Basin, Serbia** 159
Milka M. Vidovic and Vojin B. Gordanic

Chapter 6 **Conceptual Models in Hydrogeology,
Methodology and Results** 181
Teresita Betancur V., Carlos Alberto Palacio T.
and John Fernando Escobar M.

Chapter 7 **Groundwater Management by Using
Hydro-Geophysical Investigation: Case Study:
An Area Located at North Abu Zabal City** 201
Sultan Awad Sultan Araffa

Permissions

List of Contributors

Preface

This book is a great source of information regarding the field of hydrogeology primarily intended for the people interested in the fields of earth sciences, environmental sciences, and physical geography. In the past few years, the world has started paying great attention on the area of groundwater hydrology and the inculcation of hydrogeology. It is mainly due to the rising requirement to meet the acute shortage of water, especially the increasing demand of ground water. This book has been written by a number of analysts belonging to different countries. It clearly explains a wide area of problems in the field of hydrogeology. Karst hydrogeology and removal procedures, hydrogeochemistry, soil hydraulic properties as a component influencing groundwater refill procedures, purposeful theoretical structures, and geophysical exploration for groundwater have all been described in this book. It provides the reader with an outlook on the work of hydrogeologists and scientists on the optimization of groundwater resources. This book will prove to be immensely beneficial to graduates and workers of this field.

Significant researches are present in this book. Intensive efforts have been employed by authors to make this book an outstanding discourse. This book contains the enlightening chapters which have been written on the basis of significant researches done by the experts.

Finally, I would also like to thank all the members involved in this book for being a team and meeting all the deadlines for the submission of their respective works. I would also like to thank my friends and family for being supportive in my efforts.

Editor

Hydrogeology of Karstic Area

Haji Karimi
Ilam University
Iran

1. Introduction

Karst is a special type of landscape that is formed by the dissolution of soluble rocks. Karst regions contain aquifers that are capable of providing large supplies of water. More than 25 percent of the world's population either lives on or obtains its water from karst aquifers. In the United States, 20 percent of the land surface is karst and 40 percent of the groundwater used for drinking comes from karst aquifers. Natural features of the landscape such as caves and springs are typical of karst regions. Karst landscapes are often spectacularly scenic areas.

2. Karst definition and different types of karst

The term karst represents terrains with complex geological features and specific hydrogeological characteristics. The karst terrains are composed of soluble rocks, including limestone, dolomite, gypsum, halite, and conglomerates. As a result of rock solubility and various geological processes operating during geological time, a number of phenomena and landscapes were formed that gave the unique, specific characteristics to the terrain defined by this term. Karst is frequently characterized by karrens, dolines (sinkholes), shafts, poljes, caves, ponors (swallowholes), caverns, estavelles, intermittent springs, submarine springs, lost rivers, dry river valleys, intermittently inundated poljes, underground river systems, denuded rocky hills, karst plains, and collapses. It is difficult to give a very concise definition of the word karst because it is the result of numerous processes that occur in various soluble rocks and under diverse geological and climatic conditions (Milanovic, 2004).

The main features of the karst system are illustrated in Figure 1. The primary division is into erosional and depositional zones. In the erosional zone there is net removal of the karst rocks, by dissolution alone and by dissolution serving as the trigger mechanism for other processes. Some redeposition of the eroded rock occurs in the zone, mostly in the form of precipitates, but this is transient. In the net deposition zone, which is chiefly offshore or on marginal (inter- and supratidal) flats, new karst rocks are created. Many of these rocks display evidence of transient episodes of dissolution within them (e.g. Alsharhan and Kendall, 2003).

Within the net erosion zone, dissolution along groundwater flow paths is the diagnostic characteristic of karst. Most groundwater in the majority of karst systems is of meteoric origin, circulating at comparatively shallow depth and with short residence time underground. Deep circulating, heated waters or waters originating in igneous rocks or subsiding sedimentary basins mix with the meteoric waters in many regions, and dominate the karstic dissolution system in a small proportion of them. At the coast, mixing between seawater and fresh water can be an important agent of accelerated dissolution (Ford and Williams, 2007).

In the erosion zone most dissolution occurs at or near the bedrock surface where it is manifested as surface karst landforms. In a general systems framework most surface karst forms can be assigned to input, throughput or output roles. Input landforms predominate. They discharge water into the underground and their morphology differs distinctly from landforms created by fluvial or glacial processes because of this function. Some distinctive valleys and flat-floored depressions termed poljes convey water across a belt of karst (and sometimes other rocks) at the surface and so serve in a throughput role (Ford and Williams, 2007).

Some karsts are buried by later consolidated rocks and are inert, i.e. they are hydrologically decoupled from the contemporary system. These are referred as palaeokarsts. They have often experienced tectonic subsidence and frequently lie unconformably beneath clastic cover rocks. Contrasting with these are relict karsts, which survive within the contemporary system but are removed from the situation in which they were developed, just as river terraces – representing floodplains of the past – are now remote from the river that formed them. Relict karsts have often been subject to a major change in baselevel. A high-level corrosion surface with residual hills now located far above the modern water table is one example; drowned karst on the coast another. Drained upper level passages in multilevel cave systems are found in perhaps the majority of karsts (Ford and Williams, 2007).

Karst rocks such as gypsum; anhydrite and salt are so soluble that they have comparatively little exposure at the Earth's surface in net erosion zones, in spite of their widespread occurrence. Instead, less soluble or insoluble cover strata such as shales protect them. Despite this protection, circulating waters are able to attack them and selectively remove them over large areas, even where they are buried as deeply as 1000 m. The phenomenon is termed interstratal karstification and may be manifested by collapse or subsidence structures in the overlying rocks or at the surface. Interstratal karstification occurs in carbonate rocks also, but is of less significance. Intrastratal karstification refers to the preferential dissolution of a particular bed or other unit within a sequence of soluble rocks, e.g. a gypsum bed in a dolomite formation (Ford and Williams, 2007).

Cryptokarst refers to karst forms developed beneath a blanket of permeable sediments such as soil, till, periglacial deposits and residual clays. Karst barre´ denotes an isolated karst that is impounded by impermeable rocks. Stripe karst is a barre´ subtype where a narrow band of limestone, etc., crops out in a dominantly clastic sequence, usually with a stratal dip that is very steep or vertical. Recently there has been an emphasis on contact karst, where water flowing from adjoining insoluble terrains creates exceptionally high densities or large sizes of landforms along the geological contact with the soluble strata (Kranjc, 2001).

Karst-like landforms produced by processes other than dissolution or corrosion-induced subsidence and collapse are known as pseudokarst. Caves in glaciers are pseudokarst, because their development in ice involves a change in phase, not dissolution. Thermokarst is a related term applied to topographic depressions resulting from thawing of ground ice. Vulcanokarst comprises tubular caves within lava flows plus mechanical collapses of the roof into them. Piping is the mechanical washout of conduits in gravels, soils, loess, etc., plus associated collapse. On the other hand, dissolution forms such as karren on outcrops of quartzite, granite and basalt are karst features, despite their occurrence on lithologies of that are of low solubility when compared with typical karst rocks (Ford and Williams, 2007).

When there is also a sufficient hydraulic gradient, this can give rise to turbulent flow capable of flushing the detached grains and enlarging conduits by a combination of mechanical erosion and further dissolution. Thus in some quartzite terrains vadose caves develop along the flanks of escarpments or gorges where hydraulic gradients are high. The same process leads to the unclogging of embryonic passages along scarps in sandy or argillaceous limestones. Development of a phreatic zone with significant water storage and permanent water-filled caves is generally precluded. The landforms and drainage characteristics of these siliceous rocks thus can be regarded as a style of fluvio-karst, i.e., a landscape and subterranean hydrology that develops as a consequence of the operation of both dissolution and mechanical erosion by running water (Ford and Williams, 2007).

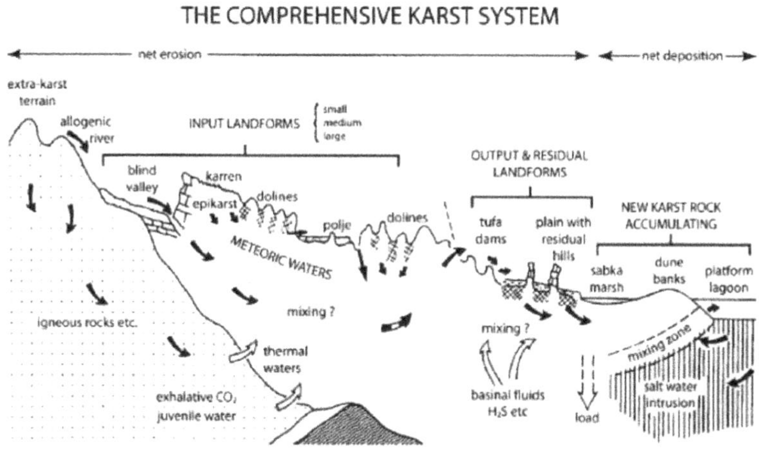

THE COMPREHENSIVE KARST SYSTEM

Fig. 1. The comprehensive karst system: a composite diagram illustrating the major phenomena encountered in active karst terrains (Ford and Williams, 2007).

3. Surface features of karst terrains

Since the beginning of karst studies is the surface geology, the surface karst features are the signature of karst performance in the area. Distinguishing and recognition of these phenomenons denote to the development of karst. Different karst features like various types of karrens, dolines (sinkholes), ponors, poljes and springs will introduce and their mechanism of formation will be discussed.

3.1 Karrens

The characteristics of karrens are mainly adopted from Gunn (2004). Limestone that outcrops over large areas as bare and rocky surfaces is furrowed and pitted by characteristic sculpturing landforms that generate a distinctive karstic landscape. These solutional forms, ranging in size from less than 1 mm to more than 30 m, are collectively called karren, an anglicized version of the old German word Karren (the equivalent of the French terms lapiés and lapiaz). Currently, these groups of complex karren forms tend to be called karrenfields or Karrenfelder, in order to differentiate such large-scale exokarst landforms from their smaller karren components (see Table 1).

Several different weathering processes may produce microkarren over limestone surfaces. Some of the microkarren features, such as biokarstic borings, are the result of specific solutional processes induced by cyanobacteria, fungi, algal coatings, and lichens.

At this scale, many different patterns of minute hollows and pits are common, especially in arid environments, because the occasional wetting of the rock produces irregular etching, frequently coupled with biokarstic action. Microrills are the smallest karren form showing a distinctive rilling appearance. Microrills consist of very tiny and sinuous runnels, 0.5-1 mm wide, rarely more than 5 cm long; they are caused by dew and thin water films, enhanced in coastal locations by supralittoral spray. Some other specific karren features develop near the coastline.

The majority of etched surfaces in semiarid environments display a rather complex microtopography that rarely presents linear patterns, the only exception being microrills. The general trend is a chaotic and holey limestone surface in which focused corrosion dominates, without any kind of integration in drainage patterns. These solutional features related to focused corrosion, give rise to depressions of different sizes, more or less circular in plan, such as the rainpit and the kamenitza karren types. Rainpits are small cup-like hollows, sub-circular in plan and nearly parabolic in cross section, whose diameter ranges from 0.5-5 cm and rarely exceed 2 cm in depth; they appear clustered in groups, or even packed by coalescence. The kamenitza karren type (Table 1) consists of solution pans, generally flat-bottomed, from a few square centimeters to several square meters in size, that are produced by the solutional action of still water that accumulates after rainfall; their borders, frequently elliptical or circular in plan, are overhanging and may have small outlet channels.

Many types of karren are linear in form, controlled by the direction of channeled waters flowing along the slope under the effect of gravity. The smaller ones are called rillenkarren and are easy to distinguish from solution runnels or rinnenkarren by their trough width, which rarely exceeds 4 cm. Rillenkarren can be defined as narrow solution flutes, closely packed, less than 2.5 cm in mean width, consisting of straight grooves separated by sharp parallel ribs, that are initiated at the rock edges and disappear downwards. Rillenkarren are produced by direct rainfall and their limited extent seems to be explained by the increase of water depth attaining a critical value that inhibits further rill growth downslope. Neither dendritic patterns nor tributary channels can be recognized in rillenkarren flutes, as opposed to the normal (Hortonian) erosional rills.

Solution runnels are not as straight and regular in form as rillenkarren, being greater and more diversified in shape and origin. Solution runnels or rinnenkarren are normal (Hortonian) rills and develop where threads of runoff water are collected into channels. Classification of solution runnels is difficult because of the great diversity of topographic conditions, the complex processes involved, and the specific kind of water supply feeding the channel. Rinnenkarren is the common term to describe the equivalent of Horton's first-order rills on soluble rocks; they result from the breakdown of surface sheetflows that concentrate into a channelled way and they are also wider than rillenkarren. These solution runnels are sculpted by the water runoff pouring down the flanks of the rocks and have distinctive sharp rims separating the channels; their width and depth range from 5–50 cm, being very variable in length (commonly from 1–10 m, but in some cases exceeding 20 m long). Rundkarren are rounded solution runnels developed under soil cover; they differ from rinnenkarren in the roundness of the rims between troughs and can be considered good indicators of formerly soil-covered karren. Many transitional types from rundkarren to rinnenkarren can be found, due to deforestation and re-shaping of the rocks after subsequent soil removal by erosion. Undercut runnels or hohlkarren are associated with semi-covered conditions, as suggested by the bag-like cross sections of the channel, resulting from enhanced corrosion at the soil contact. Decantation runnels are rills, which reduce in width and depth downslope because the solvent supply is not directly related to rainfall, but corresponds to overspilling stores of water, such as moss clumps, small snow banks, or soil remnants. Wall karren are the typical straight runnel forms developing on sub-vertical slopes, but meandering runnels are more frequent on moderately inclined surfaces or where some kind of decantation feeding occurs over flat areas or gentle slopes. Wall karren may attain remarkable dimensions exceeding 30 m in length. Obviously, transitional forms of runnels are abundant in the majority of karren outcrops, with the exception of areas with arid climates.

Other types of karren features are linear forms controlled by fractures. Grikes or kluftkarren are solutionally widened joints or fissures, whose widths range from 10 cm to 1 m, being deeper than 0.5 m and several meters long. Grikes are one of the commonest and widespread karren features and separate limestone blocks into tabular intervening pieces, called clints in the British literature and Flachkarren in German. For this reason, clint and grike topography is the most typical trend in the limestone pavements, such as the Burren (Ireland; see separate entry) and Ingleborough (northwest England; see Yorkshire Dales entry). The term "cutters" is commonly used in North America as a synonym for grike, although it is best applied to a variety of grike that develops beneath soil cover. Giant grikes, larger than 2 m wide to over 30 m deep, are called bogaz or corridors. Corridor karst or labyrinth karst constitutes the greatest expression of this type of fracture-controlled karrenfield. Splitkarren are similar smaller scale features, resulting from solution of very small weakness planes, being less than 1 cm deep and 10 cm long. Since they conduct water to the karst aquifers, grikes are very important.

Finally, there is a group of karren features closely related to the solutional action of unchannelled washing by water sheets. Many of them, particularly trittkarren and solution ripples, show a characteristic trend that is transverse to the rock slope. At the foot of rillenkarren exposures, subhorizontal belts of unchannelled surfaces can be observed; they are called solution bevels and appear as smoothed areas flattened by sheet water corrosion. More distinctive forms are trittkarren or heelsteps, which are the result of complex solutional processes involving both horizontal and headward corrosion resulting from the thinning of water sheets flowing upon a slope fall. The single trittkarren consists of a flat tread-like surface, 10–40 cm in diameter, and a sharp backslope or riser, 3–30 cm in height.

A wide variety of peculiar karren forms are produced by special conditions, such as where solution takes place in contact with snow patches or damp soil. Trichterkarren are funnel-shaped forms that resemble trittkarren, but are formed at the foot of steep outcrops where

Solutional agent	Karren forms							Synonyms
Biokarstic	Borings							
Weting		Irregular etching						
Tiny water films		Microrills						Rillensteine
Storm shower			Rainpits					Solution pits
Direct rainfall			Rillenkarren					Solution flutes
Chanelled water flow					Solution runnels			Rinnenkarren
						Wall karren		Wandkarren
					Decantation runnels			
					Meandering runnels			Maanderkarren
Standing water				Kamenitza				Solution pans
Sheet wash water flow				Solution bevels				Ausgleichflachen
				Trittkarren				Heelsteps
		Cockling patterns						
			Solution ripples					
Snow melting				Trichterkarren				Funnel karren
				Sharpened edges				Lame dentates
					Decantation runnels			
					Meandering runnels			Maanderkarren
Iced melting						Meandering runnels		
Infiltration					Grikes			Kluftkarren
Soil percolation water					Rundkarren			Rounded runnels
				Smoth surfaces				Bodened runnel,
				Subsoil tubes				Subcataneous karren
				Subsoil hollows				
					Cutters			

Complex processes	0-1mm	1mm-1cm	1-10cm	10cm-1m	1-10m	10-100m	100m-1km	>1km	Lapies
					Undercut runnels				Hohlkarren
					Clints				Flachkarren
					Pinnacles				Spitzkarren
							Pinnacle karrenfield		Karrenfeld
							Limestone pavement		
								Stone forest	
								Arete karst	
	0-1mm	1mm-1cm	1-10cm	10cm-1m	1-10m	10-100m	100m-1km	>1km	Lapies

Table 1. Classification of karren forms. Light grey areas enclose elementary karst features. Dark grey areas enclose complex large-scale landforms, namely karren assemblages and karrenfield types (Gines et al., 2009)

snow accumulates. Sharpened edges or "lame dentate", as funnel karren features, are developed beneath snow cover. Rounded smooth surfaces, associated with subsoil tubes and hollows are very common subcutaneous forms, due to the slow solution produced in contact with aggressive water percolating through the soil.

In Bögli's classifications, two kinds of complex karren forms are recognized: clints or flachkarren, and pinnacles or spitzkarren. These latter, three-dimensional forms, range from 0.5–30 m in height and several meters wide, and are formed by assemblages of single karren rock features, being the constituents of larger-scale groups of complex forms, the karrenfields or karrenfelder. Pinnacles or spitzkarren are pyramidal blocks characterized by sharp edges, resulting from the solutional removal of rock from their sides, as well as from cutting through furrow karren features. Pinnacles are exceptionally well developed in the tropics, where spectacular landscapes constituted by very steep ridges and spikes have been reported. In some cases, such as the Shilin or Stone Forest of Lunan, the presence of transitional forms, evolving from subsoil dissected stone pinnacles sometimes called "dragons' teeth" to huge and rilled pinnacles more than 30 m in height, can be observed.

Karrenfields are bare, or partly bare, extensions of karren features, from a few hectares to a few hundred square kilometres. Additional work is needed to clarify the relation between karren assemblages and climate, on the basis of the current knowledge accumulated in the last decades from arctic, alpine, humid-temperate, mediterranean, semiarid, and humid-intertropical karsts.

3.2 Sinkhole

Sinkholes are "enclosed hollow of moderate dimensions" originating due to dissolution of underlying bedrock (Monroe, 1970). More specially, sinkholes are surficial landform, found in karst areas and consist of an internally drained topographic depression that is generally circular, or elliptical in plain view, with typically bowel, funnel, or cylindrical shape. Although the circular plan view and funnel shape are ideal forms for sinkholes, they may coalesce into irregular groups or have shapes that are much more complex (Wilson, 1995). The terms sinkholes and dolines are synonymous.

Sinkholes develop by a cluster of inter-related processes, including bedrock dissolution, rock collapse, soil down-washing and soil collapse. Any one or more of these processes can create a sinkhole. The basic classification of sinkholes has six main types that relate to the dominant process behind the development of each, the main characteristics of which are shown in Table 2 and further considered below.

From the lowest point on their rim, their depths are typically in the range of a few meters to tens of meters, although some can be more than a hundred meters deep and occasionally even 500 m. Their sides range from gently sloping to vertical, and their overall form can range from saucer-shaped to conical or even cylindrical. Their lowest point is often near their centre, but can be close to their rim. Dolines are especially common in terrains underlain by carbonate rocks, and are widespread on evaporite rocks. Some are also found in siliceous rocks such as quartzite. Dolines have long been considered a diagnostic landform of karst, but this is only partly true. Where there are dolines there is certainly karst, but karst can also be developed subsurface in the hydrogeological network even when no dolines are found on the surface.

The term sinkhole is sometimes used to refer both to dolines (especially in North America and in the engineering literature) and to depressions where streams sink underground, which in Europe are described by separate terms (including ponor, swallow hole, and stream-sink). Thus the terms doline and sinkhole are not strictly synonymous. Hence, to avoid the ambiguity that sometimes arises in general usage, further qualification is required, such as solution sinkhole or collapse sinkhole. Indeed, the international terminology that is used to refer to dolines that are formed in different ways can also be very confusing. Table 3 lists the terms employed by different authors, the range of terms partly reflecting the extent to which genetic types are subdivided.

The followings are the description of six main types of sinkholes which is described by Waltham and Fookes (2005):

Dissolution sinkholes are formed by slow dissolutional lowering of the limestone outcrop or rockhead, aided by undermining and small-scale collapse. They are normal features of a karst terrain that have evolved over geological timescales, and the larger features are major landforms. An old feature, maybe 1000 m across and 10 m deep, must still have fissured and potentially unstable rock mass somewhere beneath its lowest point. Comparable dissolution features are potholes and shafts, but these are formed at discrete stream sinks and swallow holes, whereas the conical sinkholes are formed largely by disseminated percolation water.

Collapse sinkholes are formed by instant or progressive failure and collapse of the limestone roof over a large cavern or over a group of smaller caves. Intact limestone is strong, and large-scale cavern collapse is rare. Though large collapse sinkholes are not common, small-scale collapse contributes to surface and rockhead degradation in karst, and there is a continuum of morphologies between the collapse and dissolution sinkhole types.

Caprock sinkholes are comparable to collapse sinkholes, except that there is undermining and collapse of an insoluble caprock over a karstic cavity in underlying limestone. They occur only in terrains of palaeokarst or interstratal karst with major caves in a buried limestone, and may therefore be features of an insoluble rock outcrop (Thomas, 1974).

Dropout sinkholes are formed in cohesive soil cover, where percolating rainwater has washed the soil into stable fissures and caves in the underlying limestone (Table 2). Rapid failure of the ground surface occurs when the soil collapses into a void that has been slowly enlarging and stoping upwards while soil was washed into the limestone fissures beneath (Drumm et al, 1990; Tharp, 1999; Karimi and Taheri, 2010). They are also known as cover collapse sinkholes.

Suffusion sinkholes are formed in non-cohesive soil cover, where percolating rainwater has washed the soil into stable fissures and caves in the underlying limestone. Slow subsidence of the ground surface occurs as the soil slumps and settles in its upper layers while it is removed from below by washing into the underlying limestone - the process of suffusion; a sinkhole may take years to evolve in granular sand. They are also known as cover subsidence sinkholes. A continuum of processes and morphologies exists between the dropout and suffusion sinkholes, which form at varying rates in soils ranging from cohesive clays to non-cohesive sands. Both processes may occur sequentially at the same site in changing rainfall and flow conditions, and the dropout process may be regarded as very rapid suffusion. Dropout and suffusion sinkholes are commonly and sensibly described collectively as subsidence sinkholes and form the main sinkhole hazard in civil engineering (Waltham, 1989; Beck and Sinclair, 1986; Newton, 1987). Subsidence sinkholes are also known as cover sinkholes, alluvial sinkholes, ravelling sinkholes or shakeholes.

Buried sinkholes occur where ancient dissolution or collapse sinkholes are filled with soil, debris or sediment due to a change of environment. Surface subsidence may then occur due to compaction of the soil fill, and may be aggravated where some of the soil is washed out at depth (Bezuidenhout and Enslin, 1970; Brink, 1984). Buried sinkholes constitute an extreme form of rockhead relief, and may deprive foundations of stable footings; they may be isolated features or components of a pinnacled rockhead. They include filled sinkholes, soil-filled pipes and small breccia pipes that have no surface expression.

3.3 Polje

Geologically speaking, a polje is a large, karstic, closed depression with a flat bottom often slightly tilted towards the drainage point and surrounded by steep walls and prone to intermittent flooding (Gams, 1978; Prohic et al., 1998). Poljes tend to be areas used for settlement and economic development; they are often the only arable areas in karstic regions where bare rock outcrops predominate with no soil formation. In this sense, polje flooding is poorly understood and requires greater study in order to mitigate its socioeconomic impact. The first step towards taking preventive measures against this phenomenon should be to establish the dynamics and to determine the cause of the flooding, which may be an unusual high supply of surface water and/or groundwater (Lopez-Chicano et al., 2002).

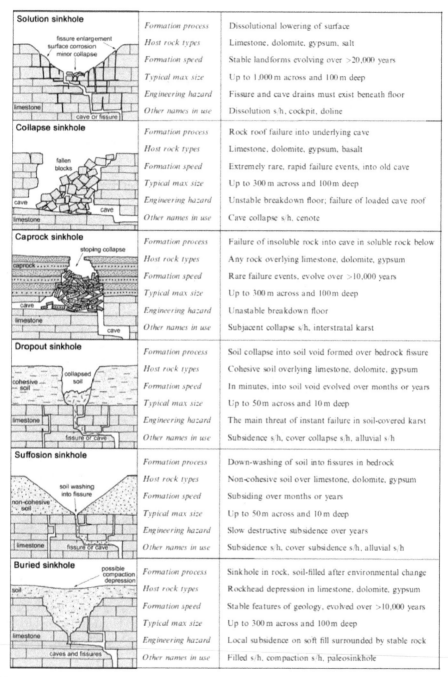

Solution sinkhole		
	Formation process	Dissolutional lowering of surface
	Host rock types	Limestone, dolomite, gypsum, salt
	Formation speed	Stable landforms evolving over >20,000 years
	Typical max size	Up to 1,000 m across and 100 m deep
	Engineering hazard	Fissure and cave drains must exist beneath floor
	Other names in use	Dissolution s/h, cockpit, doline
Collapse sinkhole		
	Formation process	Rock roof failure into underlying cave
	Host rock types	Limestone, dolomite, gypsum, basalt
	Formation speed	Extremely rare, rapid failure events, into old cave
	Typical max size	Up to 300 m across and 100 m deep
	Engineering hazard	Unstable breakdown floor; failure of loaded cave roof
	Other names in use	Cave collapse s/h, cenote
Caprock sinkhole		
	Formation process	Failure of insoluble rock into cave in soluble rock below
	Host rock types	Any rock overlying limestone, dolomite, gypsum
	Formation speed	Rare failure events, evolve over >10,000 years
	Typical max size	Up to 300 m across and 100 m deep
	Engineering hazard	Unstable breakdown floor
	Other names in use	Subjacent collapse s/h, interstratal karst
Dropout sinkhole		
	Formation process	Soil collapse into soil void formed over bedrock fissure
	Host rock types	Cohesive soil overlying limestone, dolomite, gypsum
	Formation speed	In minutes, into soil void evolved over months or years
	Typical max size	Up to 50 m across and 10 m deep
	Engineering hazard	The main threat of instant failure in soil-covered karst
	Other names in use	Subsidence s/h, cover collapse s/h, alluvial s/h
Suffosion sinkhole		
	Formation process	Down-washing of soil into fissures in bedrock
	Host rock types	Non-cohesive soil over limestone, dolomite, gypsum
	Formation speed	Subsiding over months or years
	Typical max size	Up to 50 m across and 10 m deep
	Engineering hazard	Slow destructive subsidence over years
	Other names in use	Subsidence s/h, cover subsidence s/h, alluvial s/h
Buried sinkhole		
	Formation process	Sinkhole in rock, soil-filled after environmental change
	Host rock types	Rockhead depression in limestone, dolomite, gypsum
	Formation speed	Stable features of geology, evolved over >10,000 years
	Typical max size	Up to 300 m across and 100 m deep
	Engineering hazard	Local subsidence on soft fill surrounded by stable rock
	Other names in use	Filled s/h, compaction s/h, paleosinkhole

Table 2. The six types of sinkholes, with typical cross sections and major parameters for each type (Waltham et al., 2005).

Doline-forming processes	Ford and Williams (1989)	White (1988)	Jennings (1985)	Bogli (1980)	Sweeting (1972)	Culshaw and Waltham (1987)	Beck and Sinclair (1986)	Other terms in use
Dissolution	solution	solution	solution	solution	solution	solution	solution	
Collapse	collapse	collapse	collapse	Collapse (fast) or subsidence (slow)	collapse	collapse	collapse	
Caprock collapse			Subjacent collapse		Suffusion subsidence			Interstratal collapse
Dropout	Subsidence	Cover collapse	subsidence	alluvial	alluvial	Subsidence	Cover collapse	
Suffusion	suffusion	Cover subsidence					Cover subsidence	Raveled, shakehole
Burial								Filled, paleosubsidence

Table 3. Doline/sinkhole English language nomenclature as used by various authors (modified from Waltham and Fookes, 2002).

For a depression to be classified as a polje, Gams (1978) identified three criteria that must be met:

1. Flat floor in rock (which can also be terraced) or in unconsolidated sediments such as alluvium;
2. A closed basin with a steeply rising marginal slope at least on one side;
3. Karstic drainage.

He also suggested that the flat floor should be at least 400 m wide, but this is arbitrary because Cvijic´ (1893) took 1km as a lower limit. In fact, poljes vary considerably in size. The floors of reported poljes range from ~1 to > 470 km² in area (Lika Polje is the largest at 474 km²), but even in the Dinaric karst most are less than 50 km², and elsewhere in the world a majority are less than 10 km² (Ford and Williams, 2007).

Ford and Williams (2007) categorized polje to the three basic types namely border, structural and baselevel poljes.

3.4 Ponor

Concentrated inflows of water from allogenic sources sink underground at swallow holes (also known as swallets, stream-sinks or ponors). They are of two main types: vertical point-inputs from perforated overlying beds and lateral point-inputs from adjacent impervious rocks. The flow may come from: (i) a retreating overlying caprock, (ii) the updip margin of a stratigraphically lower impermeable formation that is tilted, or (iii) an impermeable rock across a fault boundary. A perforated impermeable caprock will funnel water into the karst in much the same way as solution dolines, except that the recharge point is likely to be defined more precisely and the peak inflow larger. Inputs of this kind favour the development of large shafts beneath. Lateral-point inputs are usually much greater in volume, often being derived from large catchments, and are commonly associated with major river caves. The capacity of many ponors in the Dinaric karst exceeds 10m³/s and the capacity of the largest in Biograd-Nevesinjko polje is more than 100m³/s (Milanovic, 1993). When the capacity of the swallow hole is exceeded, back-flooding occurs and surface overflow may result (Ford and Williams, 2007).

3.5 Caves

The definition adopted by most dictionaries and by the International Speleological Union is that a cave is a natural underground opening in rock that is large enough for human entry. This definition has merit because investigators can obtain direct information only from such caves, but it is not a genetic definition. Ford and Williams (2007) define a karst cave as an opening enlarged by dissolution to a diameter sufficient for 'breakthrough' kinetic rates to apply if the hydrodynamic setting will permit them. Normally, this means a conduit greater than 5–15mm in diameter or width, the effective minimum aperture required to cross the threshold from laminar to turbulent flow.

Isolated caves are voids that are not and were not connected to any water input or output points by conduits of these minimum dimensions. Such non-integrated caves range from vugs to, possibly, some of the large rooms occasionally encountered in mining and drilling. Protocaves extend from an input or an output point and may connect them, but are not yet enlarged to cave dimensions.

Where a conduit of breakthrough diameter or greater extends continuously between the input points and output points of a karst rock it constitutes an integrated cave system. Most enterable caves are portions of such systems (Ford and Williams, 2007).

Culver and White (2005) present a general cave classification and Palmer (1991) presents a genetic classification of caves.

3.6 Springs

Karst springs are those places where karst groundwater emerges at the surface. Karst spring discharge ranges over seven orders of magnitude, from seeps of a few milliliters per second to large springs with average flows exceeding 20 m³/s. Flow may be steady, seasonal, periodic, or intermittent, and may even reverse. Karst springs are predominantly found at low topographic positions, such as valley floors, although they may be concealed beneath alluvium, rivers, lakes, or the sea (vrulja). Some karst springs emerge at more elevated positions, usually as a result of geological or geomorphological controls on their position (Gunn, 2004).

Springs in non-karst rocks may result from the convergence of flow in a topographic depression or from the concentration of flow along open fractures such as faults, joints, or bedding planes. Flow in porous media is limited by hydraulic conductivity, so that associated springs almost always have very small flow, often discharging over an extensive "seepage face." Larger springs are possible in fractured rocks such as basalt, where flow may be concentrated along open or weathered fractures. What distinguishes karst springs is that they are the output points from a dendritic network of conduits, and therefore tend to be both larger and more variable in discharge and quality than springs arising from coarse granular or fractured media.

In general, karst springs can be considered in terms of their hydrological function, their geological position, and their karstic drainage or "plumbing". Karst springs have been classified in many different ways. In theory, different attributes could be combined to describe a spring. For instance the spring at Sof Omar Cave, Ethiopia could be described as a "perennial, full-flow, gravity resurgence". In practice, most karst springs are described in terms of their most important attribute, depending upon the interest of the observer and the context of the application (Gunn, 2004).

The location and form of karst springs is determined primarily by the distribution of karst rocks, and the pattern of potential flow paths (fractures) in the rock (Karimi et al., 2005). Where karst rocks are intermixed with impermeable rocks, the latter act as barriers to groundwater flow, and karst springs tend to develop as "contact springs" where the boundary between the karst and impermeable rock is exposed at the surface. Where the impermeable unit underlies the karst, it enhances the elevation of the karst water, and the spring (and aquifer) is considered "perched", as it lies above the topographically optimum discharge point. Where the impermeable unit overlies the karst aquifer, it enhances the pressure of karst water, and springs are then described as "confined perched springs, and so exhibit more sustained flow (Gunn, 2004).

The quality and magnitude of flow from a karst spring reflects the form and function of the karst aquifer, and in particular the recharge processes and the conduit network. Springs deriving much of their water from allogenic surface catchments are known as resurgences. Springs in autogenic aquifers, which receive the bulk of their recharge from a karst surface, are known as exsurgences and they exhibit less variability in discharge and composition. In the past, such flow behaviour has been attributed to distinctive "diffuse", "conduit" and "pseudo-diffuse" (Karimi et al., 2003) Karst aquifers, but it is now recognized that recharge or underflow-overflow effects are responsible, and that a diffuse karst aquifer is an oxymoron (Gunn, 2004).

A few karst springs show remarkable periodicity in their flow, with a typical period of minutes to hours. In general, this is attributed to the existence of an internal siphon, which progressively fills and drains. Periodicity in hydrothermal springs is seen in geysers. The key feature of geysers is the warming of a pressurized body of water to boiling point and the explosive spontaneous boiling occurring as pressure is released.

Many karst springs occur adjacent to or beneath the surface of rivers, lakes, or the sea; the majority is likely unacknowledged. The interaction between the aquifer and the external water body rests on the hydraulic head distribution and the pattern of connections (springs, sinks, and estavelles) that exists.

Where karst spring water is supersaturated, calcareous tufa deposits develop at the orifice and downstream. Such petrifying springs mantle all objects in calcite, and often build up distinctive mounds and barrages in areas of peak precipitation.

3.6.1 Spring hydrograph analysis

Karst-spring hydrograph analysis is important, first, because the form of the output discharge provides an insight into the characteristics of the aquifer from which it flows and, second, because prediction of spring flow is essential for careful water resources management. However, although the different shapes of outflow hydrographs reflect the variable responses of aquifers to recharge, Jeannin and Sauter (1998) expressed the opinion that inferences about the structure of karst systems and classification of their aquifers is not efficiently accomplished by hydrograph analysis because hydrograph form is too strongly related to the frequency of rainfall events. If a long time-series of such records is represented as a curve showing the cumulative percentage of time occupied by flows of different magnitude, then abrupt changes of slope are sometimes revealed in the curve, which have been interpreted by Iurkiewicz and Mangin (1993), in the case of Romanian springs, as representing water withdrawn from different parts of the karst system under different states of flow. For these reasons, analysis of the recession limbs of spring hydrographs offers considerable potential insight into the nature and operation of karst drainage systems (Bonacci 1993), as well as providing information on the volume of water held in storage. Sauter (1992), Jeannin and Sauter (1998) and Dewandel et al. (2003) provide important recent reviews of karst-spring hydrograph and chemograph analyses (Ford and Williams, 2007).

The principal influence on the shape of the output hydrograph of karst springs is precipitation. Rain of a particular intensity and duration provides a unique template of an input signal of a given strength and pattern that is transmitted in a form modified by the aquifer to the spring. The frequency of rainstorms, their volume and the storage in the system, determines whether or not recharge waves have time to pass completely through the system or start to accumulate. Antecedent conditions of storage strongly influence the proportion of the rainfall input that runs off and the lag between the input event and the output response. The output pattern of spring hydrographs is, however, moderated by the effect of basin characteristics such as size and slope, style of recharge, drainage network density, geological variability, vegetation and soil. As a consequence of all the above, flood hydrograph form and recession characteristics show considerable variety (Ford and Williams, 2007).

Given widespread recharge from a precipitation event over a karst basin, the output spring will show important discharge responses, characterized by:

1. A lag time before response occurs;
2. A rate of rise to peak output (the 'rising limb');
3. A rate of recession as spring discharge returns towards its pre-storm outflow (the 'falling limb');
4. Small perturbations or 'bumps' on either limb although best seen on the recession.

When the hydrograph is at its peak, storage in the karst system is at its maximum, and after a long period of recession storage is at its minimum. The slope of the subsequent recession curve indicates the rate of withdrawal of water from storage. The characterization of the rate of recession and its prediction during drought are necessary for determining storage and reserves of water that might be exploited.

Maillet's exponent implies that there is a linear relation between hydraulic head and flow rate (commonly found in karst at baseflow), and the curve can be represented as a straight line with slope -α if plotted as a semilogarithmic graph. It can be represented in logarithmic form as

$$\log Q_t = \log Q_0 - 0.4343t\alpha \qquad (1)$$

from which α may be evaluated as:0

$$\alpha = \frac{\log Q_1 - \log Q_2}{0.4343\,(t2 - t1)} \tag{2}$$

Semi-logarithmic plots of karst spring recession data often reveal two or more segments, at least one of which is usually linear (Figure 2). In these cases the data can be described by using separate expressions for the different segments. Jeannin and Sauter (1998) and Dewandel et al. (2003) explain the various models that have been used to try to conceptualize the structure of the karst drainage system that has given rise to the hydrograph form observed and the means by which its recession might be analysed. If the karst system is represented as consisting of several parallel reservoirs all contributing to the discharge of the spring and each with its individual hydraulic characteristics, then the complex recession of two or more linear segments can be expressed by a multiple exponential reservoir model:

$$Q_t = Q_{01}e^{-\alpha 1t} + Q_{02}e^{-\alpha 2t} + \cdots + Q_{0n}e^{-\alpha nt} \tag{3}$$

Milanovic (1976) interpreted the data for the Ombla regime (Figure 2) in Croatia as indicating flow from three types of porosity, represented by the three recession coefficients of successive orders of magnitude. He suggested that α_1 is a reflection of rapid outflow from caves and channels, the large volume of water that filled these conduits emptying in about 7 days. Coefficient α_2 was interpreted as characterizing the outflow of a system of well-integrated karstified fissures, the drainage of which lasts about 13 days; and α_3 was considered to be a response to the drainage of water from pores and narrow fissures including that in rocks, the epikarst and soils above the water table, as well as from sand and clay deposits in caves.

Bonacci (1993) provides a discussion of various causes for changes in the value of recession coefficients.

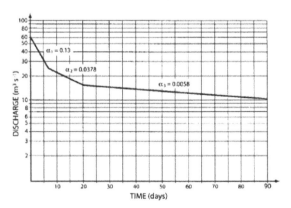

Fig. 2. Composite hydrograph recession of Ombla spring, Croatia (Ford and Williams, 2007).

3.6.2 Quality of karst spring waters

The water emerging from a karst spring consists of a mixture of water from various recharge routes and storage zones. As the environment and duration of recharge, and storage vary, so too will the resulting composition of the water. For example, allogenic recharge water will tend to be more turbid and chemically dilute than autogenic recharge. Long-term storage may result in depletion of the dissolved oxygen in the water, and deep flow may lead to warming or mineralization. In principle, these natural tracers should allow the source and routing of karst spring water to be derived. However, many of these characteristics (e.g. temperature, turbidity, dissolved oxygen, hardness) do not have fixed values associated with particular environments, they are not conserved in transit, and mixing with other waters may induce chemical reactions. The chemical composition of spring waters often results in distinctive deposits, biota, and exploitation, allowing a chemical classification (Gunn, 2004).

3.6.3 Calculation of catchment area and determination of the boundary

The catchment area of each spring is estimated by the following equation:

$$A=Q/(P.I) \tag{4}$$

In which A is the catchment area of the spring (km^2), Q is the total annual volume of water discharging from the spring [million cubic meters (MCM)], P is the total annual precipitation (m) and I is the recharge coefficient (%).

Considering the discharge variations of a spring, one of the following cases is possible:

1. Spring discharge is more or less the same at the beginning and end of the water balance year. In this condition, there are relatively no changes in the system storage and the mean annual discharge of a spring can be used for calculating the annual volume of discharge.
2. If the discharges of a spring at the beginning and end of the water balance year are different; the total discharge of the spring due to the water balance year's precipitations could be estimated by the method proposed by Raeisi (2008).

Determination of recharge coefficient is very difficult. A large number of factors like existence of sinkholes, density of joints and fractures, their openings and type and extent of infillings, percentage, thickness and granulation of soil cover, slope of beds and topography, depth, type, time and space distribution of precipitation, temperature, vegetation cover, etc. can affect the recharge coefficient. Recharge coefficient is defined as the percentage of precipitation, which contributes to the groundwater (spring water). In order to determine the catchment area, it is necessary to have a good approximation of the recharge coefficient for the study area. If there is no idea about the recharge coefficient, the catchment area could be calculated by different recharge coefficients and it will be verified by the proposed method (discussed in the next section) for determining the boundary of a catchment area. Based on experiences in Zagros Mountain Ranges of Iran (Water Resources Investigation and Planning Bureau, 1993; Rahnemaaie, 1994; Raeisi, 1999; Karimi et al., 2001) recharge coefficients can vary between 40 to 90 percent of precipitation. The lower limit is related to areas of low precipitation, high temperature and evaporation and thick soil coverage. The upper limit represents the existence of sinkholes and well-developed karst features.

One of the most complex and difficult problems to deal with in karst hydrology, hydrogeology and geology is the determination of exact catchment boundaries and area of the springs and streamflows (Bonacci and Zivaljevic, 1993). The determination of the catchment boundaries and the catchment area is the starting point in all hydrologic analyses and one of the essential data, which serve as a basis for all hydrologic calculations. In order to exactly define the surface and subsurface catchment boundaries, it is necessary to conduct detailed geologic investigations and, accordingly, extensive hydrogeologic measurements. These measurements primarily involve connections (links) between individual points in the catchment area (connections: ponors-springs, piezometers-piezometers, piezometers-springs) applying one of the tracing methods constituting dye tests, chemical tests, solid floating particles or radioactive matter. The catchment areas in karst vary according to the groundwater levels that change with time. Only in exceptional cases do the surface and subsurface watershed lines coincide and only in those places where the boundaries between catchments are located in impermeable rocks. If this boundary is located in permeable carbonate layers it is not stable.

Figure 3 shows three cases outlining the relationship between a topographic (orographic) and hydrologic (hydrogeologic) spring catchments in the karst. In most cases the basic topographic catchment area A_t is smaller than the hydrologic area A_h, whose boundaries are located within the hydrologic catchment as shown in figure 3A. In practice, it is easy to determine the topographic catchment area, whereas the determination of the hydrologic catchment area is a complex task, difficult to carry out precisely and reliably.

It should be primarily stressed that the definition of the exact catchment area and of the position of the exact catchment boundaries in karst is an interdisciplinary task. It can be exactly and completely carried out only with very close cooperation between various scientists, geologists, hydrogeologists and hydrologists, not excluding the collaboration with the researchers engaged in other scientific disciplines (Karimi, 2003).

It is especially important to define the position of the underground catchment boundaries in karst, and in the analysis of their changes, which is related to the groundwater levels. In order to carry out this task properly it is necessary to install a certain number of piezometers, plan their position and optimum number and to monitor continually the changes of groundwater level in the catchment. In the first phase a small number of piezometers should be installed, and later new ones should be added according to the analysis of the groundwater levels oscillations in the piezometers installed in the first phase.

Estimation of the catchment area in karst can be treated as an inverse problem from the hydrological standpoint. The input vectors are the elements of the water budget (rainfall, inflows, evapotranspiration). Knowing the output vectors (spring discharge or inflow into the river section in karst) it is possible to determine the catchment area, which its position and size ensure the optimum agreement between the hydrograph obtained by calculations and the hydrograph, defined by measurements (Bonacci, 1987, 1990; Bonacci and Zivaljevic, 1993). A promising method in estimation of catchment area is based on the analysis of the groundwater hydrograph, which is caused mainly by groundwater discharge fluctuations. The hydrograph variations, which are dependent on groundwater inflows, describe the outflow from one groundwater reservoir to an open spring. This method can be applied to hydrograph analysis of springs in general and especially to those in karst.

However, according to the basinwide hydrological budget calculations (Degirmenci and Gunay, 1993; Cardillo-Rivera, 2000), hydrological balance and dye tracing tests (Forti et. al,

1990, Karimi et al., 2005), the subsurface catchments were found to be considerably larger than the hydrographic basins

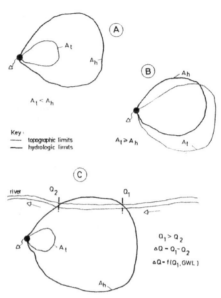

Fig. 3. Three relations between topographic A_t and hydrologic A_h catchment area for karst springs. (A) A_t; (B) A_h; (C) special case when a permanent streamflow is included in the spring catchment along one section (Bonacci, 1987)

Using the calculated approximate catchment area, the most probable location and boundaries of the catchments can be determined by the following procedure:

Step 1.   All limestone in the anticline related to the spring and the neighbouring anticlines with higher elevations than the spring is considered as the catchment area.

Step 2.   In the area determined in step 1, there must be no hydrogeological and tectonic barriers disconnecting the hydrogeological relationship between the karst aquifer and the spring; In other words, geological and tectonic settings justify the catchment area. Exact and very good geological cross sections could be very useful in this stage. Areas with hydrogeological barriers were disregarded from the catchment area.

Step 3.   A general water balance is considered for the area determined in the step 2; i.e. all the outputs (including discharge to alluvium) and also all the inputs will be taken into account, so that the catchment area of the spring does not interfere with the other springs.

The catchment area is probably as close as possible to the spring, i.e. at first, the catchment area is supplied by the spring related anticline.

Step 4.   The following parameters are useful in confirming of the determined area:

1.   The physico-chemical parameters of the spring display the characteristics of the related karst aquifer and adjacent formations.

2. Isotope studies can be used quantitatively or qualitatively in the determination of the elevation ranges that have a main role in the recharging of the spring.
3. The general direction of flow may be determined using water table elevations or isopotential maps, if piezometers are constructed in the study area.
4. Flow coefficient (FC) parameters is a good representations of the springs which are recharging from the outside the surface catchment area (FC≥ 1). It is defined as the ratio of volume of discharge to the volume of precipitation on the surface catchment area of a given hydrometric station.
5. The normalized base flow (NBF) diagram, which is a useful tool, is used for checking the calculated catchment areas (Connair and Murray, 2002; White, 2002).
6. The characteristics of the catchments such as soil cover, vegetation, sinkholes, morphology, amount and type of precipitation and elevation could be applied to differentiate between different catchment areas.

Step 5. Tracing and geophysics: The catchment areas determined with a high uncertainty about them, could be verified using tracing and geophysical tests. Because of the uncertainty in recharge coefficient, the error in the determined catchment area could be as high as 10 percent.

The catchment area of the Alvand Basin springs, west of Iran, was calculated based on equation 4, and according to the above criteria the most probable boundary of the catchment area of the main karst springs was determined (Figure 4).

3.6.4 Time variations of physico-chemical parameters

Temporal variations of physico-chemical parameters of karst springs have been used to determine aquifer characteristics in the last four decades. Jakucs (1959) showed that the chemical composition and discharge of karst spring water might vary with time. Garrels and Christ (1965) categorized the flow in karst regions into open and closed systems, based on the amounts of CO_2 available for dissolution. White and Schmidt (1966) and White (1969) classified the flow in karst aquifers into conduit and diffuse flow. Karimi et al. (2003) introduced pseudo diffuse flow regime. Shuster and White (1971, 1972) concluded that the type of flow (diffuse or conduit) can be determined by its chemograph. Ternan (1972), Jacobson and Langmuir (1974), Ede (1972), Cowell and Ford (1983) evaluated the flow systems in karst formations using other criteria, such as type of recharge, electrical conductivity and coefficient of variation of electrical conductivity, coefficient of variation of discharge and temperature, and time variations of these parameters.

Based on the above ideas, in a diffuse flow system, recharge is generally conducted through a network of numerous small joints and fractures that are distributed in the karst aquifer. The openings of these fractures are smaller than one centimetre and water slowly reaches the groundwater in a laminar manner. One of the main peculiarities of these aquifers is the small variation of physical and chemical properties of the discharging springs. Natural discharge from such a system is usually through a large number of smaller springs and seeps. In a conduit flow system, the aquifer is fed through either large open fractures (ranging from one centimetre to more than one meter) or sinkholes. In such systems, water reaches the groundwater very quickly and ultimately the springs in a turbulent manner. Hence, the physico-chemical properties of the spring waters are non-uniform. In this type of system, the discharge usually occurs through one single large spring.

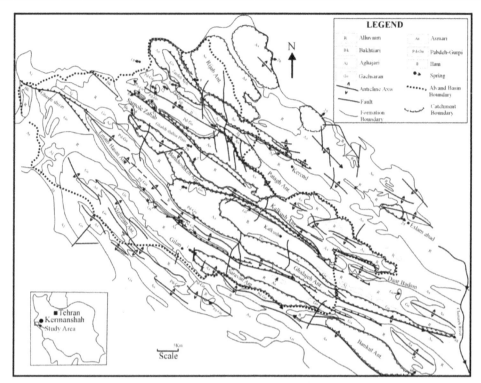

Fig. 4. Boundary of catchment area of the Alvand main karst springs (Karimi et al., 2005).

Bakalowicz (1977), Atkinson (1977), Scanlon and Thrailkill (1987), and Raeisi et al. (1993) were not able to use the criteria proposed by previous workers to determine the flow regime and found contradictory results. The reason is probably due to the fact, that purely diffuse or purely conduit flow systems rarely occur in nature, rather it is a combination of these two types of flow that usually prevail. Raeisi and Karami (1996) suggested that when the physico-chemical characteristics of a karst spring are to be used to determine the properties of the related aquifer, the first step should be the evaluation of the effects of external factors on the outflow. Lopez-Chicano et al. (2001) analyzed the hydrogeo-chemical processes in waters of Betic Cordilleras in Spain by studying hydrography, temporal evolution of physico-chemical parameters, ionic ratios (mainly Mg/Ca) and by means of simple and multivariate statistical analysis. They concluded that the aquifer exhibits diffuse flow.

Time series variations of physico-chemical parameters of springs were inspected by different researchers like Ashton (1966); Hess and White (1988, 1993); Bakalowicz et al. (1974); White (1988, 2002); Williams (1983); Scanlon and Thrailkill (1987); Scanlon (1989); Sauter (1992); Ryan and Meiman (1996); Raeisi and Karami (1996, 1997); Lopez-Chicano et al. (2001) and Desmarais and Rojstaczer (2002). Generally in a typical karst system, after an intense rainfall, discharge increases within a short period and then decreases slowly. In this period, the EC shows an increasing-decreasing-increasing trend. Based on electrical conductivity response, Desmarais and Rojstaczer (2002) divided spring response into three stages. The three stages include flushing, dilution, and recovery.

Flushing: The flushing stage marks the initial response in the spring to storms. The beginning of this stage is signalled by the increase of the slope of the conductivity curve of the spring. There are two hypotheses concerning the water source that causes this flushing of the spring:

1. The flushed water is water that has interacted or equilibrated within the soil zone, and possibly resides in small pores or fractures near the land surface, i.e. the subcutaneous zone. This water would be relatively warm and would likely contain dissolved salts (or would dissolve salts from the soil during transit), which would give the water a relatively high conductivity. The warmer, high electrical conductivity water would be mobilized by the rainwater infiltrating into the soil and pushed toward the spring.

2. The new rainwater is able to rapidly recharge the aquifer, possibly through fractures or surface swallets, and it mobilizes older, deeper water that has been residing in smaller fractures and pores out of the aquifer. This 'old water' is at or near equilibrium with the limestone, but the new water is not. The old water, because it has resided in the aquifer for a relatively long time, would have higher electrical conductivity than the baseflow spring water. Flushing is not typical of all carbonate springs (Ryan and Meiman, 1996; Desmarais and Rojstaczer, 2002).

Dilution: The dilution phase begins with the peak in the electrical conductivity (EC) curve and ends when the EC reaches its minimum value. The start of the conductivity decrease represents the first arrival of storm water at the spring. During this phase, the temperature commonly levels off and then remains constant until the next storm whereas discharge continues to decrease. The area of the recharge basin is the main factor controlling the length of this phase. After that time period, the spring begins to 'recover' because there is very little recharge water remaining.

However, the system response can also be explained by a competition between the velocity at which recharge water is moving through the system, how fast this 'new' water dissolves carbonates to gain the same chemistry signature as the 'old' aquifer water, and the amount of mixing that takes place between these two water sources.

Recovery: The recovery phase begins when the minimum is obtained in conductivity. During this phase, conductivity increases steadily until the next storm begins. The concentrations of all the major cations and anions increase during this period. All of these changes indicate that the system is returning to equilibrium conditions. The conductivity minimum likely indicates that the last of the recharge water has been in contact with the aquifer rock long enough to begin to dissolve limestone and/or dolomite in sufficient quantities to allow the overall system to begin to recover from the dilution. Figure 5 shows the three above-mentioned stages in the SS-5 spring in Bear Creek Valley, Tennessee (Desmarais and Rojstaczer, 2002).

The minimum of electrical conductivity corresponds to the maximum dilution of groundwater by fresh recharged water, which could be used as a representative of lag time of the system. The lag time is a measure of the length of time required for the arrival of unsaturated water (minimum of EC) to reach to the recording station (Hess and White, 1988).

Hess and White (1993) stated that fluctuation in hardness fit the well-established concept that hardness variability is an indication of conduit karstic drainage system as has been observed for many conduit karst aquifers in North America and Europe (Pitty, 1966; Ternan, 1972; Atkinson, 1977).

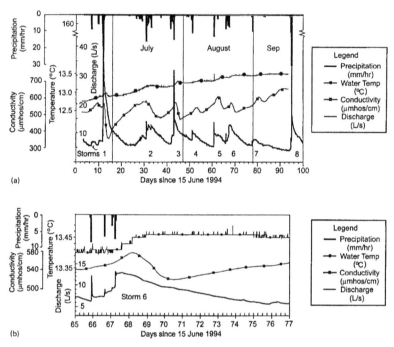

(a)

(b)

Fig. 5. (a) Discharge, conductivity and temperature at SS-5 for storms 1-8. (b) Detailed Discharge, conductivity and temperature at SS-5 for storm 6 (Desmarais and Rojstaczer, 2002).

Conductivity measurements of the spring water provide an inexpensive and rapid method of distinguishing the rock type through which groundwater flows. The amount of variability of the data through time also gives insight to how rapidly the water quality changes by recharge events. Springs with high conductivity and low coefficient of variation of the data suggest slow groundwater movement through a non-karst aquifer. The data with great variability indicates that groundwater flow is rapid through conduits (Ogden et al., 1993).

3.6.5 Environmental isotopes in hydrogeology

Environmental isotopes now routinely contribute to hydrogeological investigations, and complement geochemistry and physical hydrogeology. Meteoric processes modify the stable isotopic composition of water, and so the recharge waters in a particular environment will have a characteristic isotopic signature. This signature then serves as a natural tracer for the provenance of groundwater. On the other hand, radioisotopes decay, providing us with a measure of circulation time, and thus groundwater renewability. Environmental isotopes provide, however, much more than indications of groundwater provenance and age. Looking at isotopes in water, solutes and solids tells us about groundwater quality, geochemical evolution, recharge processes, rock-water interaction, the origin of salinity and contamination processes (Clark and Fritz, 1997).

Given the relationship between $\delta^{18}O$ in precipitation and elevation, it is possible to determine an approximate mean recharge elevation for springs. An inherent assumption in this type of analysis is that the spring water is young enough to be comparable to modern precipitation (James et al., 2000). On the basis of the relationship established between the $\delta^{18}O$ value in rainwater and altitude, Kattan (1997); Abbott et al. (2000); James et al. (2000); Yoshimura et al. (2001); Eisenlohr et al. (1997); Vallejos et al. (1997) and Ellins (1992) estimated the mean elevation of recharge zones of groundwater in their study areas. These estimations corresponded more or less to the natural topographic divide line in their study areas.

Even at the same altitudes, there are variations in the $\delta^{18}O$ values of respective rains because stable isotope compositions of water vapour masses are different from one to another. Consequently, the mean values of $\delta^{18}O$ of rainwater for about a half or full year are plotted against the altitude (Abbott et al., 2000; Yoshimura et al., 2001). Altitude isotope effects were observed in different parts of the world. For example Kattan (1997) reported a figure of -0.23 ‰ per 100 meter increase in elevation in Syria and Abbott et al., (2000) a figure of -0.25 ‰ per 100 meter in USA (Vermont) and other researchers like James et al., (2000); Yoshimura et al. (2001); Vallejos et al. (1997) and Williams and Rodoni (1997) figures of -0.18‰, -0.15 to -0.25‰, -0.35‰ and -0.1‰ per 100 m increase in elevation in USA (Oregon), a tropical area, Spain and California respectively. Leontiadis et al. (1996) provided an estimate of -0.44‰ 100 m⁻¹ for the groundwater altitude effect on the $\delta^{18}O$ value in Eastern Macedonia and -0.21‰ in Thrace in Greece. Figure 6 shows the determination of recharge elevations from the relationship between snow $\delta^{18}O$ and elevation. Approximate recharge elevations can be determined by extrapolating from the spring isotopic composition to the regression line, then dropping a perpendicular line to intersect the abscissa. Inset is the precipitation data used to constrain the regression line (James et al., 2000).

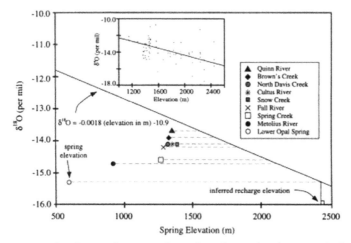

Fig. 6. Determination of recharge elevations from the relationship between $\delta^{18}O$ in snow and elevation (James et al., 2000).

Stewart and Williams (1981) stated that the large seasonal variations in $\delta^{18}O$ values in the recharge zones diminish to within experimental error at the springs due to diffusion in the very large groundwater reservoir, thus little information on the rate of groundwater flow can be obtained from these data. Ferdrickson and Criss (1999) found that the isotope variations for the Meramec River Basin (Missouri) springs share the overall, cycloid-like shape of the precipitation, with higher $\delta^{18}O$ values during the summer and fall and excursions to more negative values during the winter months. The $\delta^{18}O$ variations for the precipitation have amplitude exceeding 10‰, yet the annual amplitude of the variations in the rivers was only 3‰ and the amplitudes of several karst springs were even smaller, valuing at about 1‰ (Figures 1-10 and 1-11). They used these seasonal variations to determine the residence time of water.

Isotope studies indicate that there is some dilution in spring water during the rain season because of the mixing of event water with the spring and the enrichment in the dry season (Stewart and Williams, 1981; Ford and Williams, 1989; Ferdrickson and Criss, 1999; Abbot and Bierman, 2000).

4. Different zones of a karst aquifer

Generally an unconfined aquifer can be subdivided into two saturated and unsaturated zones. It is not necessary that all parts of this classification be present in any given karst (Karimi, 2003).

In the unsaturated zone above the water table, voids in the rock are only partially occupied by water, except after heavy rainfall when some voids fill up completely. Water is percolated downward in this zone by a multiphase process, air and water co-existing in the pores and fissures. Because of the high concentration of CO_2 in the soil cover, the percolating water in the soil cover is usually unsaturated and aggressive. When water leaves the soil cover, it behaves as in a closed system because of the lack of any connection to the atmosphere (Ford and Williams, 1989). The subcutaneous zone is situated under the soil cover or in the upper part of the limestone aquifer.

The water-saturated zone below the water table is the phreatic zone (White, 1988). If the karst mass is large and deep enough, it is possible to distinguish a shallow and a deep phreatic zone. The circulation of groundwater is fast in the former and slow in the latter. Different scientists estimated the depth of the shallow phreatic zone to be between 20 to 60 m (Bogli, 1980).

The lower boundary of an aquifer is commonly an underlying impervious formation. But should the karst rocks be very thick, the effective lower limit of the aquifer occurs where no significant porosity has developed. Figure 7 shows alternative approaches to estimating the thickness of an unconfined karst aquifer.

5. Problems of construction in karst regions

Karst processes and landforms pose many different problems for construction and other economic development. Every nation with karst rocks has its share of embarrassing failures such as collapse of buildings or construction of reservoirs that never held water. Prevention of unanticipated remedial measures in karst terrains now imposes a lot of economic problems each year on the governments.

5.1 Rock slide-avalanche hazards in karst

A landslide or rock slide-avalanche is the catastrophically rapid fall or slide of large masses of fragmented bedrock such as limestone (Cruden, 1985). 'Landslide' is more widely used but is also applied to slides of unconsolidated rocks. Rock slides take place at penetrative discontinuities, the mechanical engineering term for any kind of surface of failure within a mass. Once initiated, there is powerful momentum transfer within the falling mass and it may partially ride on a cushion of compressed air that can permit it to run for some hundreds of metres upslope on the other side of a valley (van Gassen and Cruden, 1989).

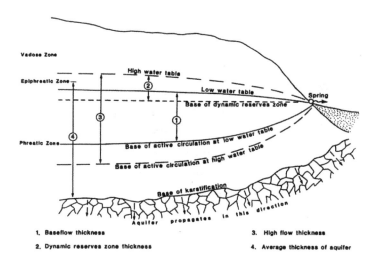

Fig. 7. Alternative approaches to estimating the thickness of an unconfined karst aquifer (Ford and Williams, 1989).

Carbonate rocks and gypsum are especially failure-prone for two reasons.

1. While faults and joints are the only important penetrative discontinuities in most other rocks, in karst strata there is also major penetration via bedding planes. In fact they are particularly favoured as surfaces of failure because of their great extent.
2. Large quantities of water may pass rapidly through the rock via its karst cavities to saturate or lubricate interlaminated or underlying weak or impermeable strata such as clays. The forces that resist catastrophic failure within a particular rock are defined by an internal angle of friction. Minimum angles for relatively hard carbonates without shale interbeds range from 14° to 32°.

The principal settings of landslides in karst rocks are shown in Figure 8. Slab slides are particularly common because they are bedding-plane failures. They are especially frequent and dangerous in the overdip situation. The biggest one in the world is the Saymareh landslide in western Iran (Karimi, 2010). Rotational failures within massive carbonates are comparatively rare but there are large ones in dolomites in the Mackenzie Mountains, Canada. Toppling cliffs are common in all rocks; see Cruden (1989) for formal analysis. Toppling or rotational failures are quite common along escarpment fronts where the permeable karst rock rests on a weak but impermeable base such as a shale; Ali (2005) describes 12 limestone failures of up to 800×10^6 t each along a 20 km frontage near the city of Sulaimaniya in northern Iraq, caused by spring sapping at the contact with underlying shales.

Downslope detachment and creep of karst rock formations resting on slick but impermeable strata beneath them (Figure 8) may proceed slowly for long periods and then suddenly accelerate into a landslide, usually as a consequence of heavy rains or an earthquake. At the Vajont Dam disaster of 1963 in the Italian Alps, in which 2000 lives were lost; rise of water level in a reservoir may have contributed by increasing pore water pressures on the slide plane. The Ok Ma Landslide (Papua New Guinea) was a slide of $\sim 36 \times 10^6$ m^3 of fractured massive limestone on clay dipping into a river valley that was induced by removal of the toe of a previous slide in order to install a dam for a gold mine.

5.2 Setting foundations for buildings, bridges, etc.

Setting foundations where there are soils, etc., covering maturely dissected epikarst can encounter many problems. Figure 9 illustrates the range of different methods that are used to overcome them by compacting the soil or pinning the footings to (comparatively) firm bedrock. Under large or heavy structures the majority of these methods can be very expensive. Reinforced concrete slabs ('rafts' floating on the soil following its mechanical compaction) are now much used as alternatives under buildings. For roads on mantled epikarst or spanning infilled solution and suffusion dolines strong synthetic plastic sheeting, strips or meshes ('geofabrics') are being substituted because they are cheaper: their longterm reliability is not yet established, however. Much has been written on these subjects; see Beck (2005) and Waltham et al. (2005) for recent surveys.

Building calamities remain frequent worldwide.

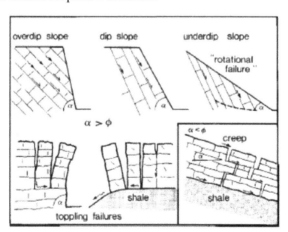

Fig. 8. Types of landslides (or rock slide-avalanches) in carbonate rocks. Ø is the internal angle of friction of the rock. Failures on dip and overdip slopes are termed 'slab slides' (Ford and Williams, 2007).

Cavities entirely within bedrock can also pose dangers if they are at very shallow depth or if the planned structural load is considerable. For typical strong limestones with caves, Waltham et al. (2005) recommend a minimum of 3m bedrock above a cavity 5m wide, 7m for widths >10m; for chalk and gypsum, at least 5m of rock above a cave 5m wide.

Construction on gypsum requires particular care. Gutierrez (1996) and Gutierrez and Cooper (2002) discuss the rich example of Calatayud, a town of 17000 persons in the Ebro Valley, Spain. It is built on a fan of gypsiferous silts interfingering with floodplain alluvium, and underlain by a main gypsum formation ~500m thick. The existing buildings are 12th century to modern in age. Many (of all ages including recent) display subsidence damage that ranges from minor to very severe. The primary cause is believed to be dissolution of the gypsum bedrock, which is abetted by local compaction of overburden accumulated since the town was founded in AD 716, collapse of some abandoned cellars and dissolution of the silts.

5.3 Tunnels and mines in karst rocks

Tunnels and mine galleries (adits or levels) will be cut through rocks in one of three hydrogeological conditions:

(i) vadose; (ii) phreatic but at shallow depth or where discharge is limited, so that the tunnel serves as a transient drain that permanently draws down the water table along its course; (iii) phreatic, as a steady-state drain, i.e. permanently water-filled unless steps are taken drain it. Long tunnels in mountainous country may start in the vadose zone at each end but pass into a transient zone, or even a steady-state phreatic zone, in their central parts.

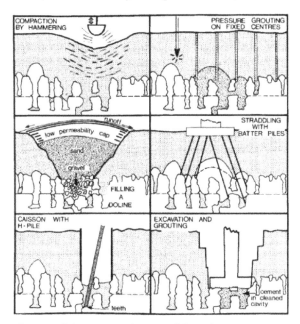

Fig. 9. Illustrations of some of the principal types of foundation treatments in a soil-mantled karst (Ford and Williams, 2007).

Vadose and transient zone tunnels are cut on gentle inclines to permit them to drain gravitationally.

Where the tunnel or mine is a deep transient drain or is in the steady-state phreatic zone, gravitational drainage will not suffice, e.g. if the tunnel is below sea level. Three alternative strategies can then be adopted. The first is to pump from the tunnel itself, when necessary. It is prone to failure if the pumps fail and to disaster (for the miners) if large water-filled cavities are intercepted, causing catastrophic inrushes of water. The second means is to grout the tunnel and then to pump any residual leakage as necessary. It is the essential method for transportation tunnels. Traditionally, tunnel surfaces were rendered impermeable by applying a sealant (e.g. concrete) as they became exposed. This does not deal with the catastrophic inrush problem.

Modern practice is to drill a 360° array of grouting holes forward horizontally, then blast out and seal a section of tunnel inside this completed grout curtain. This largely deals with the hazard of catastrophic inrush, i.e. a flooded cavity should be first encountered by a narrow bore drill hole that can be sealed off quickly. Milanovic (2000) and Marinos (2005) discuss tunnel protection thoroughly, with many examples.

Grouting is not feasible in the extracting galleries of a mine. Here, a third and most elaborate strategy is to dewater the mine zone entirely, i.e. maintain a cone of depression about it for as long as the mine is worked.

A largely debated issue is related to the engineering aspects of karst. As population density increases, need for construction of roads, various infrastructures, and water resources increases. This leads to reclamation projects with construction of dams and reservoirs in or nearby karst regions. The understanding and evaluation of environmental impacts of such human activities on karst are important to try and find a balance between development and preservation of these complex hydrogeological systems (Milanovic, 2002, 2004).

While the problems associated with the construction of a dam site on a karst area are fairly well understood (Uromeihy, 2000; Romanov et al., 2003, 2007; Turkmen, 2003; Xu and Yan, 2004; Ghobadi et al., 2005) consequences of flooding of karst discharge areas due to reservoirs built close to karst regions are much less studied (De Waele, 2008).

5.4 Dam construction on carbonate rocks

A large number of dams have been built on karstic limestones and dolomites for different purposes all over the world. Flood control has been particularly important on branches of the Mississippi, where the Tennessee Valley Authority (TVA) was very active in the first half of the 20th Century. Storage to sustain paddy fields, counter prolonged dry seasons or general drought was more important in China, around the Mediterranean and in Iraq, Iran and other semi-arid areas. Hydroelectric power generation was an early priority in alpine sites and is now the principal goal of perhaps the majority of the larger, higher dams. Few nations that constructed them escaped serious problems due to karst leakage, leading to considerable overruns in cost or to outright abandonment in some instances. These are summarized in many engineering design and construction reports. The TVA main report (1949) is still pertinent; Soderberg (1979) gave a more recent review of their work. Therond (1972), Mijatovic (1981), Nicod (1997) and Milanovic´ (2000, 2004) have discussed European experience, which generally has been with geologically more complex mountainous sites.

Therond (1972) identified seven different major factors that may contribute to the general problem. These are: type of lithology, type of geological structure, extent of fracturing, nature and extent of karstification, physiography, hydrogeological situation and the type of dam to be built. For each factor, clearly, there are a number of significantly different conditions. In Therond's estimation, these together yield a combination of 7680 distinct situations that could arise at dam sites on carbonate rocks! It follows that dam design, exploration and construction must be specific to the particular site, and be continually re-evaluated:

Milanovic´ (2000) suggests that there have been three principal settings for dams on carbonate rocks:

1. In the narrow gorges typically created where large allogenic rivers cross them in steep channels. Here rates of river entrenchment have usually been faster than karst development; with the result that karstification is not a major problem beneath the channels. It may however be hazardous in the gorge walls, which will form the dam abutments.

2. Dams and reservoirs in broader valleys where the karst evolution has been as fast as or faster than river entrenchment. The TVA sites are examples. This can cause many problems beneath the dam as well as in the abutments and upstream in valley sides and bottoms. It is particularly hazardous where the valley is hanging at its mouth, as is common in alpine topography, because the natural (pre-dam) groundwater gradient is steepened there. Unfortunately, this will also be an optimum site for hydroelectric power dam location because the reservoir volume, fall height and gradient of the penstock are all maximized there.

3. In poljes to control flooding and store water for dry season irrigation. This is perhaps the most difficult setting because under natural conditions the dry season water table will be deep below the polje floor in highly karsted rock. The reservoir floor must be sealed (with clay, shotcrete, PVC, etc.) to retain water but the seals can be blown by air pressure as floodwaters rise in the caves underneath. Ponors must be plugged or walled off by individual dams rising above the reservoir water surface, and estavelles must be fitted with one-way valve systems. Much success has been achieved in the poljes of former Yugoslavia but the Cernic´a project there and Taka Polje in Greece are two examples that were abandoned after expensive study.

Many dams are more than 100 m in height and some exceed 200 m. A first, obvious danger of dam construction is that by raising the water table to such extents, an unnaturally steep hydraulic gradient is created with unnatural rapidity across the foundation and abutment rock, and an unnaturally large supply of water is then provided that may follow this gradient. This is a hazardous undertaking because, unless grout curtains penetrate well into unkarstified rock, the increased pressure will drive groundwater movement under the dam and stimulate dissolution. Dreybrodt et al. (2002, 2005) have approached this problem with realistic modeling scenarios for limestone and gypsum. In the limestone case solution conduits are shown to propagate to breakthrough dimensions (turbulent flow) beneath a 100m deep grout curtain under a dam in approximately 80 years. Remedial work would then be essential. Table 4 provides details of leakage from dams in karst before and after remedial works, and Figure 10 shows one example of increasing leakage over time at a dam in Macedonia.

While leakage through dam foundations and abutments is most feared, it is quite possible that there may be lateral leakage elsewhere in a reservoir. Problems with karst can arise even where the dam itself is built on some other rock, if karst rocks are inundated upstream of it. Montjaques Dam, Spain, was built to inundate a polje. It failed by leaking through tributary passages and the scheme was abandoned (Therond, 1972).

In tackling dams on karst the first essential is drilling of exploratory boreholes (with rock core extracted for inspection), and mining of adits (galleries big enough for human entry and inspection) in the abutments. These may later be used for grouting. Surface, downhole and interhole geophysics (Milanovic´, 2000) can amplify the picture but are not in themselves sufficient because they will rarely detect smaller cavities, or even large ones below ~50m or so. Even intensive drilling and mining may be inadequate. At the Keban Dam site in Turkey, despite 36 000m of exploratory drilling and 11 km of exploratory adits, a huge cavern of over 600 000m^3 was not detected; 'expect the unexpected!' (Milanovic´, 2000).

Grout curtains are essentially dams built within the rock. 'Due to karst's hydrogeological nature, grout curtains executed in karstified rock mass are more complex and much larger than curtains in other geological formations' (Milanovic 2004). The surest principle is to grout entirely through the limestone into underlying impermeable and insoluble strata where this is possible. Curtains in abutments can also be terminated laterally against such strata (the 'bathtub' solution).

The normal practice is to excavate all epikarst and fill any large caverns discovered by the adits and bores, then place a main curtain beneath the dam, in the abutments and on the flanks. A cut-off trench and second, denser curtain may be placed upstream in the foundation if there are grave problems there. In the main curtain a first line of airtrack grout holes will be placed on centers never more than 8–10m apart and filled until there is back pressure. A second, offset line of holes is then placed and filled between them. Third and fourth lines may be used until the spacing reaches a desirable minimum that is normally not more than 2m. Adits in the abutments that are used to inject grout should be no more 50m apart vertically. Standard grouts are cement with clay (particularly bentonite, a clay that expands when wetted), plus sand and gravel for large cavities. Mixtures are made up as slurries with differing proportions of water. Ideally, the goal of all grouting is to reduce leakage of water to one Lugeon unit (Lu=1 L min^{-1}m^{-1} of hole at 10 bars water pressure) under a dam and 2 Lu in the abutments. In practice, in karsted limestones it is often difficult to inject grout where permeability is <5.0 Lu. Correlation between Lugeon measured during exploration and the amount of grout that will be required can be very poor also; at Grancarevo Dam, Herzegovina, consumption ranged from 1.5 to 1500 kgm^{-1} in different holes that had recorded only ~1.0 Lu before grouting began (Milanovic, 2000).

All springs and piezometers must be monitored carefully as the reservoir fills behind a completed dam. Operators should be prepared to halt filling and drain the reservoir as soon as serious problems appear. In extreme cases the reservoir floor and sides may be sealed off, e.g. by plastic sheeting. Experience shows that remedial measures after a dam has been completed and tested are much more costly than dense grouting during construction.

Dam/reservoir	After first filling (m³ s⁻¹)	After remedial works (m³ s⁻¹)
Keban (Turkey)	26	< 10
Camarassa (Spain)	11.2	2.6
Mavrovo (FYR Macedonia)	9.5	Considerably reduced
Great Falls (USA)	9.5	0.2
Marun (Iran)	10	Considerably reduced
Canelles (Spain)	8	Negligible
Slano (Yugoslavia) (34 m³ s⁻¹)*	8	(3.5) Increase till 6
Ataturk (Turkey)	> 11	?
Višegrad (Bosnia)	9.4	Remedial work runs
Buško Blato (Bosnia) (40 m³ s⁻¹)*	5	3
Dokan (Iraq)	6	No leakage
Contreas (Spain)	3–4	?
Hutovo (Herzegovina) (10 m³ s⁻¹)*	3	1
Gorica (Herzegovina)	2–3	No remedial works
Špilje (FYR Macedonia)	2	No remedial works
El Cajon (Honduras)	1.65	0.1
Krupac (Yugoslavia)	1.4	Negligible
Charmine (France)	0.8	0.02
Krušćica (Sklope) (Croatia)	0.8	0.35
Mornos (Greece)	0.5	Considerably reduced
Piva (Yugoslavia)	0.7–1	No remedial works
Maria Cristina (Spain)	20% of inflow	?
Peruća (Croatia)	1	No remedial works
Sichar (Spain)	20% of inflow	?
La Bolera (Spain)	0.6	?

Table 4. Leakage from reservoirs reduced after remedial works (Milanovic, 2004).

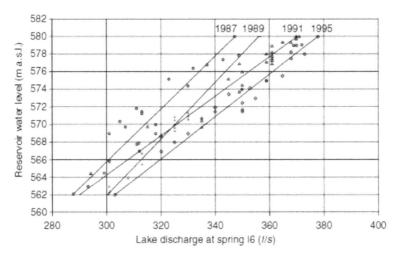

Fig. 10. Increasing leakage at Spring I6 below the Spilje Dam, Macedonia, related to water level in the reservoir behind the dam (Ford and Williams, 2007).

Despite such intense effort dams still fail to achieve design levels in karst. A good example is the Lar Dam, Elbruz Mountains, Iran. This is a 105m high earth-fill dam in a hanging valley at 2440m a.s.l. The geology is complex. The natural water table was >200m beneath the dam, draining to major springs 8 km distant and 350m lower in elevation. A sequence of international engineering firms tackled it beginning in the 1950s. During the first attempt to fill the reservoir, leakage via the springs rose to 60–80% of the inflow. It was drained and re-grouted, with 1000–40 000 kgm[-1] of grout being injected in the worst places. A cavern of >90 000m[3] was also discovered and filled. Water losses remain unacceptably high.

5.5 Dam construction on gypsum and anhydrite

Numerous case histories are provided by A. N. James (1992) that illustrate the wide range of serious difficulties that have been encountered by building dams on evaporate rocks. These include the rapid enlargement of existing conduits and the creation of new ones, because hydraulic gradients are excessive (Figure 11); the settling or collapse of foundations or abutments where gypsum is weakened by solution; heave of foundations where anhydrite is hydrated, and attack by sulphate-rich waters upon concrete in the dam itself:

$$CaCO_3 + 2H_2SO_4 = Ca^{2+} + SO_4^{2-} + CO_2 + H_2O \qquad (5)$$

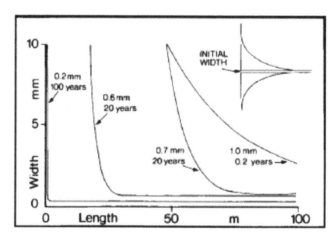

Fig. 11. Penetration distances or progress of the dissolution front for ~L99 in massive gypsum, calculated for initial fissure widths ranging 0.21–1.0 mm. Time elapsed since initiation is in years. The hydraulic gradient is 0.2 and water temperature is 10°C. (Inset) (Ford and Williams, 2007).

In their model analysis, Dreybrodt et al. (2002) obtained kinetic breakthrough beneath a 100m deep grout curtain in gypsum in 20–30 years, the conduits enlarging to give unacceptable rates of leakage within the ensuing five years.

In the USA experience has been gained in simple geological, low relief terrains in west Texas and New Mexico, where it is possible to avoid truly excessive hydraulic gradients and the problems of complex structure. At one celebrated site, McMillan Dam, gypsum is present only in the abutments, further reducing the difficulties, and no caves were detected when it was built in 1893. Nevertheless, the reservoir drained dry via caves through the left-hand abutment within 12 years. Attempts to seal off the leaking area by a cofferdam failed because new caves developed upstream of it. Between 1893 and 1942 it is estimated that 50-106m[3] of dissolution channels were created (James and Lupton, 1978).

Dams can be built successfully in gypsum terrains where relief is low and geology simple (or where there are gypsum interbeds in carbonate strata), but comprehensive grouting is necessary and an impermeable covering over all gypsum outcrops is desirable (Pechorkin, 1986). Periodic draining and regrouting will probably be needed also.

6. Water level interpretation in karst

Bonacci (1988) stressed the important role of piezometers in explaining the ground water circulation in karst. The measurements carded out on several piezometers in the Ombla catchment made further ground water analyses possible and helped to reach theoretical and practical conclusions that are important from the engineering viewpoint.

The ground water level (GWL) measurements were carried continuously in the period between January 1988 and July 1991 on 10 piezometers located in the hinterland of the Ombla Spring.

In karstified aquifers borehole data are usually difficult to relate to aquifer structure and behavior (Bakalowicz et al. 1995). However, the distribution of porosity and hydraulic conductivity from the surface to the phreatic zone can be investigated by borehole analysis and, where the number of boreholes is large, borehole tests can provide valuable information on aquifer behavior, especially in the more porous aquifers such as coral and chalk. Only rarely do boreholes intersect major active karst drains, because the areal coverage of cave passages is usually less than 1% of an aquifer and only exceptionally above 2.5% (Worthington, 1999), but when they do their hydraulic behavior may be compared to that of a spring.

More often boreholes intersect small voids with only indirect and inefficient connection to a major drainage line, in which case hydraulic behavior in the bore is very sluggish compared with that in neighboring conduits (Ford and Williams, 2007).

Bonacci and Bonacci (2000) attempted to determine the characteristics of a karst aquifer using information on ground water level (GWL) measurements in natural holes and boreholes. The majority of karst terrains in the world are still insufficiently researched from the hydrological and hydrogeological standpoint. One of the reasons is because they are situated in less developed and less wealthy parts of the world. The second but probably more important reason is the extreme heterogeneity of the karst aquifer, which causes complexity for investigation and explanation. It is very hard to obtain reliable information, parameters and general conclusions on water circulation processes. However, significant progress has been made in the investigation of karst aquifers and related problems in the last 19 years.

The differences between the four examples from the Dinaric karst given are primarily in measurement methods, i.e. the GWL monitoring procedures and in the availability and accuracy of other hydrogeological and hydrological information.

Regardless of significant differences in measurement methods, which also stipulated different elaboration levels, similar conclusions on functioning of the karst aquifer were reached. It is obvious that karst aquifers are less homogeneous than in granular media, therefore all information, especially on GWL, is valuable for their study. In karst, modelers erroneously assume that field sampling "especially from piezometers" gives an accurate spatial and time representation of the area being modeled. The result is that from the start the model is built with parameters that are not precise and have errors that accumulate in the model results. The basic problem is how to recognize and explain information contained in the GWL data. The best and probably only solution is a holistic interdisciplinary cooperation among numerous experts in the field karstology (Bonacci and Bonacci, 2000).

Bonacci (1995) interpreted the water level fluctuations in two piezometers in the Ombla spring catchment. Figure 12 presents the relationship between the simultaneous hourly GWL measured in Piezometer 8 (ordinate) and Piezometer 9 (abscissa) during a flood hydrograph at the Ombla Spring. The formation of a loop is evident during the rising and falling of the GWL, which shows that the water flow during the analysed period is non-steady. The rising period of GWL is much shorter than the falling period. The rise in GWL measured on Piezometer 9 lasts much longer, as can be seen from Figure 12, where a certain point shows the time computed from zero, i.e. the initial moment (06:00 h on 2 December 1988).

Such phenomena were recorded by BoreUi (1966) at the Bunko Blato in the Dinaric region (Bosnia and Herzegovina).

Figure 13 gives a graphical presentation of the hourly values of GWL in Piezometer 8 and the respective discharges of the Ombla Spring. GWL in Piezometer 8, as well as in other piezometers, stagnates (either remains stable or slightly increases) in the period when the hydrograph increases from 9.32 m^3/s to 54.9 m^3/s, which occurred on 8 October 1989 from 03:00 h to 14:00 h. The discharge increase at the spring lasted 11 h and in the mean time GWL in Piezometer 8 and in other piezometers was either stable or increased slowly. It should be noted that intensive rainfall started in the catchment 2 h earlier, i.e. at 01:00 h on 8 October 1989. This phenomenon can be explained by the fact that during those 11 h some large karst conduits were filled and flow under pressure was formed only in the karst conduits and caves, whereas the small karst fissures were not filled by water, so that GWL was not raised in the entire karst massif. The filling of large fissures is carried out by a rapid turbulent flow regime, whereas the small fissures are filled by a slow laminar or transitional flow regime (Bruckner at al., 1972; Gale, 1984; Lauritzen et al., 1985; Atkinson, 1986). Accordingly, it is possible to determine the volume of large caverns by integrating the hydrograph of the Ombla Spring during an 11 h period. The volume amounts to about 1.5 × 10^6 m^3. The analysis of the other hydrographs confirmed this value as an average value, and it was adopted for further analyses. It should be stressed here that such analyses should be performed only for those hydrographs that reach maximum discharges of the Ombla Spring over 80 m^3/s.

The piezometers, if carefully located, make it possible to identify the dimensions and functions of the karst underground system. Continuous measurements of GWL and discharge make it possible to reach conclusions on the water circulation in karst under different hydrologic conditions, i.e. at low, average and high water levels (Bonacci, 1995).

Healy and Cook (2002) review methods for estimating groundwater recharge that are based on knowledge of groundwater levels. Most of the discussion is devoted to the use of fluctuations in groundwater levels over time to estimate recharge. This approach is termed the water-table fluctuation (WTF) method and is applicable only to unconfined aquifers. In addition to monitoring of water levels in one or more wells or piezometers, an estimate of specific yield is required (Healy and Cook, 2002).

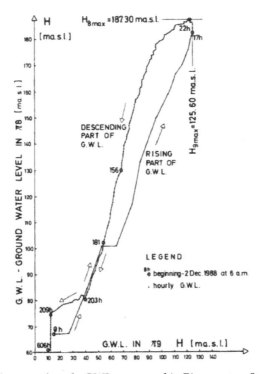

Fig. 12. Relationship between hourly GWL measured in Piezometers 8 and 9 (Bonacci, 1995).

Rehm et al. (1982) conducted a recharge study for an upland area of central North Dakota, USA. They used three different methods for estimating recharge: the water-table fluctuation method, the Hantush method, and a flow-net analysis. The 150-km²-study area contained 175 piezometers and water-table wells. The water table is in glacial deposits of sand, gravel, and loam. These deposits overlie bedrock, silt, and clay units that confine the Hagel Lignite Bed aquifer. Thirty-eight observation wells were used to estimate recharge by the WTF method. An average value for Sy of 0.16 was obtained. Estimates of recharge to the water table were made for sand and fine textured materials. Average values are shown for 1979 and 1980 in Table 5.

Ten groups of nested piezometers were used to determine vertical hydraulic gradients for the Hantush method. Hydraulic conductivity was measured in the field at each piezometer using single-hole slug tests. The measured hydraulic conductivity was assumed to be equal to the vertical hydraulic conductivity. Sites were located in the sandy and fine-textured bedrock, as well as below two of the many sloughs that are present in the study area. Measured values of K range from 3×10^{-9} m/s for clayey bedrock to 6×10^{-6} m/s for sands. Vertical gradients range from 0.006 in the sands to 1.2 in the fine-textured material. Average rates of recharge are listed in Table 5. They are considerably greater than those estimated from the WTF method. The differences are likely caused by errors inherent in the methods and natural heterogeneities within the system. Rehm et al. (1982) calculated an areal average recharge rate for the study area of 0.025–0.115 m/year by weighting the estimates in Table 5 on the basis of area coverage of the different hydrogeological settings.

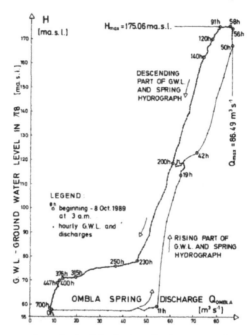

Fig. 13. Relationship between hourly Ombla Spring discharges and GWL measured in Piezometer 8 (Bonacci, 1995).

Material/location	WTF method		Hantush method	
	1979	1980	1979	1980
Sandy material	0.017	0.08	0.71	0.73
Fine-textured material	0.0018	0.0011	0.091	0.081
Sloughs	–	–	0.60	0.70

Table 5. Estimates of groundwater recharge rates, in m/year, for an upland area of North Dakota, USA, by the water-table fluctuation method (WTF) and the Hantush method for 1979 and 1980 (Rehm et al., 1982)

The basic procedure in hydraulic testing of boreholes is to inject or withdraw fluid from the hole (or a test interval within it) while measuring the hydraulic head. In the field, hydraulic conductivity is usually determined by borehole pumping tests or recharge tests (sometimes called Lugeon tests). The appropriate technique depends upon the purpose and scale of the investigation (Castany, 1984). Hydraulic conductivity may also be determined from in-hole tracer dilution (Ford and Williams, 2007).

7. Newly formed sinkholes and subsidence in karst regions

In the last two decades, by growing agricultural activities and accordingly increasing groundwater abstraction, land subsidence and sinkholes have formed in plains of different parts of the world. Various studies (Beck, 1986; Beck and Sinclair, 1986; Newton, 1987; Waltham, 1989; Benito et al., 1995; Gutierrez and Gutierrez, 1998; Tharp, 1999; Atapour and Aftabi, 2002; Salvati and Sasowsky, 2002; Abbas Nejad, 2004; Guerrero et al., 2004, 2008; Ahmadipour, 2005; Ardau et al., 2005; Gutierrez et al., 2005, 2007, 2008a; Wilson and Beck, 2005; Khorsandi and Miyata, 2007; Van Den Eeckhaut et al., 2007; Galve et al., 2008, 2009; Vigna et al., 2008; De Waele, 2009; De Waele et al., 2009; Parise et al., 2009; etc.) document problems related to sinkholes all around the world. Frequent triggering factors in the development of land subsidence and sinkholes are overexploitation from groundwater aquifers and groundwater level fluctuations. Moreover, existence of fine-grained alluvium, tectonic features especially faults and dissolution of karstic formations were reported as effective factors in the formation of sinkholes (Karimi and Taheri, 2010).

One of the most important factors in subsidence and sinkhole formation and also the rate of their formation is the water table condition and its variations. This component causes derangement in the balance between vertical stress and supporting forces in the soil column. As an example from Karimi and Taheri (2010), in order to understand the geotechnical conditions of the soil column in Famenin plain, Iran, a specific exploration borehole having 4 inches diameter, 115 m depth, rotary drilling method without drilling mud and undisturbed sampling in each 1.5 m was drilled in the distance of 15 m from Jahan Abad collapse sinkhole (6F-1). The thickness of overburden alluvium was 94 m, and the soil was mainly fine grained. The soil type from surface to the bedrock was alternation between silty clay (CL) and clayey silt (ML) with intercalations of silty and clayey sand. The soil had a medium plasticity and liquid limit ranges from 31.2 to 47.2%. The carbonate bedrock was encountered under the alluvium at a depth of 94 m, and it was composed of crashed cemented limestone fragments (Calcareous fault breccia or collapse breccia) having small voids. The void size increased at the depth range from 109 to 115 m. The water table was at a depth of 109.5 m, showing the unsaturated condition of alluvium. The borehole was sand productive, and it was implying the contribution of sand from alluvial overburden sediments.

The following factors favor the formation of sinkholes in the Famenin and Kabudar Ahang plains, Iran (Karimi and Taheri, 2010):

a. Carbonate bedrock beneath the alluvial aquifer.
b. Thick and cohesive unconsolidated soil.
c. Significant drawdown of water table due to overexploitation.
d. Penetration of deep wells into karstified limestone bedrock and predominance of turbulent flow via open joints and fractures.
e. Evacuation of alluvial aquifer materials by pumped water (sand production of wells).

Aquifer subsurface conditions and geotechnical data of 6F-1 borehole have been taken into account in order to analyze the imposed stresses into the soil column.

The assumptions of stress model on soil mass are as follows:

-   Average thickness of alluvium is 80 m.
-   The water table had been lowered 50 m during 10 years (1991–2001).

- The original water table depth was 30 m.
- The aquifer is unconfined and more or less homogenous.
- The soil is a mixture of silty clay with sand and gravel, having a dry density of 1.85 g/cm³ and a saturated density of 2.1 g/cm³.
- From water table to the bedrock, the aquifer is fully saturated.

The effective normal stress at any depth in the aquifer can be calculated with the following equation (Terzaghi et al. 1996):

$$\sigma' = (Z_1\gamma + Z_2\gamma_{Sat}) - Z_2\gamma_w \tag{6}$$

In which σ' is effective normal stress (kN/m²), Z_1 thickness above water table, Z_2 saturated thickness below water table, γ bulk density of the aquifer, γ_{Sat} saturated density of the aquifer and γ_w density of water. The water table changes create variations in the effective stresses. The imposed stress on the cavities at the bottom of the alluvial aquifer or on the top of the karst aquifer can be calculated from the above equation. Figure 14 shows a hypothetical geotechnical model and the variations of effective stress above the limestone bedrock with the mentioned assumptions. Following the decrease of water table in the area, the changes indicated below occur in the subsurface conditions:

- Loss or removal of buoyant support.
- Increase in the effective stress at the base of the soil column, for example, the effective stress increases from 1,000 to 1,400 kN/m² due to the water table drawdown.
- Natural consolidation of soil causes land subsidence.

Ford and Williams (2007) inspected the sinkhole formation by dewatering, surcharging, solution mining and other practices on karst.

7.1 Induced sinkhole formation

It is probably true to write that, after groundwater pollution, induced sinkholes are the most prominent hazardous effect of human activity in karst regions. Agriculture, mining and quarrying, highways and railways, urban and industrial constructions all contribute to the effect. Induced sinkholes generally develop more rapidly than most natural ones, appearing and enlarging in time spans ranging from seconds to a few weeks. This is faster than most societies can react with preventive or damage-limiting measures, so that such sinkholes are widely described as 'catastrophic'. Although in a few instances the hazardous collapse is of surface bedrock directly into a cave underneath, in 99% of reported cases or more it occurs in an overburden of unconsolidated cover sands, silts or clays, i.e. it is a suffusion or cover-collapse do line; 'subsidence sinkhole' is the widely used alternative term. There are two end-member processes:

1. ravelling, the grain-by-grain (or clump-by-clump) loss of detrital particles into an underlying karst cavity that is transmitted immediately by grain displacement upwards to the surface, where it appears as a funnel that gradually widens and deepens;
2. formation of a soil arch in more cohesive clays and silt–clay mixtures over the karst cavity that then stops upwards until it breaks through to the surface.

The large majority of suffusion sinkholes appear to have formed by a mixture of the two processes, with the soil above an early arch subsiding by down faulting, suffusion then sapping the fault-weakened mass to create a new arch and repeat the cycle. From the human perspective, pure soil-arch collapses (or 'dropouts') are the most dangerous because they can appear without warning at the surface. There has been much loss of life as a consequence.

The contact between a solution-indented karst surface and overlying cover deposits is the 'rockhead'. Using ground-penetrating radar (GPR) through 2–20m of cover, Wilson and Beck (1988) estimated that the frequency of karst cavities at the rockhead capable of swallowing ravelled debris varied from 12 000 to 730 000 km² in the counties of northern Florida, i.e. the rockhead there is a dense epikarst of solution pits, pipes and shafts.

Induced sinkhole formation thus has attracted a good deal of attention in recent decades. Hundreds of thousands of new subsidences have been reported worldwide, ranging from 1 × 1 × 1m holes to features >100m in length or diameter and tens of meters deep.

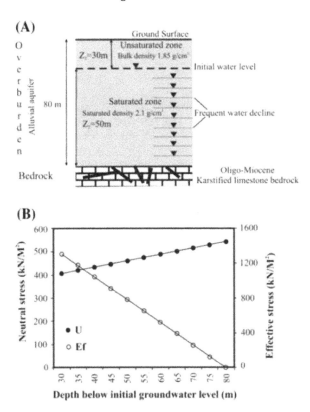

Fig. 14. A: Hypothetical geotechnical model within the soil mass, B: Relationship between effective and neutral stresses in the soil mass model (U, neutral stress; Ef, effective stress) (Karimi and Taheri, 2010).

7.2 Groundwater abstraction and dewatering

Dewatering of unconsolidated cover deposits on karst is the most important cause of induced sinkhole formation worldwide. The buoyant support of the water is removed, weakening mechanical stability. Abstraction may be for water supplies, irrigation, draining a volume of rock for mining or quarrying, or many other purposes. It is most hazardous when the cover is drained entirely so that the water table is depressed below the rockhead into the karst strata. However, sinkholes may also form readily where the lowering is limited to some level within the overburden.

Collapses are common on corrosion plains and in the bottoms of poljes when these are pumped for irrigation during dry seasons in tropical and mediterranean regions. With or without human intervention, the rapid rate of dissolution of gypsum can yield major problems. For example natural ground collapse over gypsum in northeast England resulted in about $1.5M worth of damage in the interval 1984–1994 (Cooper 1998).

Mining and quarrying tend to have the greatest impacts because they dewater to the greatest depths, usually far below the rockhead. In many coal-mining areas of China the coals are overlain by or intermingled with limestones or gypsum. Tens of thousands of sinkholes, often of large size, have been reported.

7.3 Surcharging with water

In the case of surcharging it is the addition of water at particular points that causes ravelling of overburden into karst cavities. It will therefore be especially potent where the unsurcharged water table is below the rockhead, but can also be effective with water levels in the overburden. In modern cities such point-located surcharging will be widespread unless precautions are taken, caused by drainage from individual downspouts on buildings, leaks from water supply and sewer pipes, leaking stormwater management ponds, parking lots, etc. Sinkhole formation is fastest where the overburden is thin, and is chiefly by ravelling. Sinkholes tend to be smaller than those associated with substantial dewatering, being mostly less than 10m in diameter. Nevertheless, there are many reports of building foundations being undermined and collapsed, and damage to roads and railway tracks due to neglect of soakaways or other means of dispersing stormwaters. Natural surcharging occurs on river floodplains, corrosion plains and poljes when they are inundated and is often accompanied by collapse and suffusion as the waters recede.

General rising of the water table can also create collapse or subsidence by destroying the cohesion of susceptible clay soils. However, this is comparatively rare in karst areas. The load or vibration from heavy equipment can induce small collapses locally, especially beneath it. In historic times plough-horse teams have dropped; in modern times many tractors, haulage trucks, drilling rigs and military tanks have fallen. Rock blasting from quarries or foundation cutting, etc. often causes collapses of small to intermediate scale.

7.4 Solution mining

Salt mining has induced many collapses and subsidences over the centuries. Normally this will involve significant thickness of overlying consolidated rocks. Suffusion in superficial unconsolidated deposits thus is not usually predominant, as it is in the dewatering and surcharging situations considered above. Traditionally, extraction has been by one of two methods:

1. Conventional mining via shafts and adits, removing the product by hand or machine at the workface, as in a coal mine, etc.;
2. By pumping the water from natural salt springs ('wild brine').

Recently, where feasible these have been replaced by solution mining, in which water is injected via one set of boreholes and extracted as brine via another, i.e. no workers or equipment are committed underground. The planimetric extent and volumes of cavities that will be created in the salts by both the wild brine and the injection methods are always uncertain and can be hazardous.

The most celebrated historic examples of induced subsidence in the English-speaking world are in the county of Cheshire, England, where wild brine mining began in Roman times and adit mining of salt became important with the beginning of the Industrial Revolution. Large areas of the surface have now subsided, over both open mines and solution mines, with much property damage, although catastrophically rapid collapse of ground is comparatively rare due to local geological conditions (Cooper 2001).

Corporations drilling exploratory and extractive oil wells do not expect to be involved in salt solution mining as well but this has happened in many instances in recent decades.

8. References

Atkinson, T. C. (1977) Diffuse flow and conduit flow in limestone terrain in Mendip Hills, Somerset (Great Britain). J. Hydrol., 35: 93-100.

Atkinson, T.C. (1986) Soluble rock terrain. In: P.G. Fookes and P.R. Vaughyn (Editors), Handbook of Engineering Geomorphology. Survey University Press, pp. 241-257.

Bonacci, O. (1987) Karst Hydrology. Springer, Berlin, 179 pp.

Bonacci, O. (1995) Ground water behaviour in karst: example of the Ombla Spring (Croatia), Journal of Hydrology, V. 165, p. 113-134.

Bonacc, O., Bonacci, T.R. (2000) Interpretation of groundwater level monitoring results in karst aquifers: examples from the Dinaric karst, HYDROLOGICAL PROCESSES, v 14, p. 1312-1327.

Bruckner, R.W., Hess, J.W. and White, W.B. 1972. Role of vertical shafts in the movement of ground water in carbonate aquifers. Ground Water, 10(6): 1-9.

Culver, D.C. and White, W.B. (2005) *Encyclopedia of caves*, Elsevier Academic press.

Cvijic, J., Karst: Geografska Monografija (Karst: Geographical Monograph), Beograd, Yugoslavia, 1895; first edition: Karstphanomen, Wien, 1893.

De Waele, J. (2008) Interaction between a dam site and karst springs: The case of Supramonte (Central-East Sardinia, Italy), Engineering geology, v. 99, p. 128-137.

Ford DC, Williams PW (2007) *Karst geomorphology and hydrology*, John Wiley and sons, Ltd, England, 562 p.

Gale, S.J. (1984) The hydraulics of conduit flow in carbonate aquifers. J. Hydrol., 70: 309-327.

Gams, J. (1978) The polje: the problem of definition. Z. Geomorphol. 55, 170– 181.

Gines, A., Knez, M., Slabe, T., Dreybrodt, W,.(2009) *karst rock features karren sculpturing*, ZRC Publishing, Ljubljana.

Gunn, J. (2004) *Encyclopedia of cave and karst science*, New York London, 1940 p.

Gusic, B. and Gusic, M., *Kras Carsus Iugoslaviae*, Academia Scientiarum, Zagreb, Yugoslavia, 1960.

Healy, R.W., Cook, P.G. (2002) Using groundwater levels to estimate recharge, Hydrogeology Journa, V. 10, p. 91–109.

Karimi H (2003) Hydrogeological behavior of Alvand karst aquifers, Kermanshah (in English). PhD Thesis, University of Shiraz, Iran.

Karimi, H. (2007) The formation mechanism of Zamzam well in order to enhance the engineering projects, Report, Ilam University (In Persian).

Karimi, H. (2010) The great Saymareh landslide, the agent of earthquake in the time of it occurrence, 2d nation conference on earthquake and retrofit of structure, Behbahan, Iran.

Karimi, H., E. Raeisi, M. Zare (2001). Determination of catchment area of aquifer bearing Tangab dam site using water balance method, Proceedings of The second national conference on engineering geology and the environment, Tehran, 16-18 Oct. 2001, V.2, P. 773-755.

Karimi, H.; Raeisi, E.; Zare, M. (2003) Hydrodynamic Behavior of the Gilan Karst Spring, West of the Zagros, Iran, Cave and Karst Science, V.30, No.1, P.15-22.

Karimi, H.; Raeisi E.; Bakalowicz M. (2005) Characterising the main Karst aquifers of the Alvand basin, Northwest of Zagros, iran, by a hydrogeochemical approach, Hydrogeology Journal, V. 13, Nos. 5-6, P:787 – 799.

Karimi, H., Ashjari, J. (2009) Periodic breakthrough curve of tracer dye in the Gelodareh Spring, Zagros, Iran, CAVE AND KARST SCIENCE Vol. 36, No. 1, p.: 5-10.

Karimi, H., Taheri K. (2010) Hazards and Mechanism of sinkholes on Kabudar Ahang and Famenin Plains of Hamadan, Iran, Natural Hazards, v.55, No.2, p.481-499.

Kranjc, A. (ed.) (2001) Classical Karst – Contact Karst; a Symposium. Acta Carsologica, 30(2), 13-164.

Lauritzen, S.E., Abbot, J., Arnesen, R., Crossley, G., Grepperud, D., Ive, A. and Johnson, S. (1985) Morphology and hydraulics of an active phreatic conduit. Cave Sci., 12: 139-146.

Lopez-Chicano M., Calvache M.L., Martın-Rosales W., Gisbert, J. (2002) Conditioning factors in flooding of karstic poljes—the case of the Zafarraya polje (South Spain), Catena 49: 331– 352.

Mijatovic, B.F. (1984) Karst poljes in Dinarides, in Hydrogeology of the Dinaric Karst (ed. B.F. Mijatovic), International Contributions to Hydrogeology 4, Heise, Hannover, pp. 87– 109.

Milanovic, P.T. (1993) The karst environment and some consequences of reclamation projects, in Proceedings of the International Symposium on Water Resources in Karst with Special Emphasis on Arid and Semi Arid Zones (ed. A. Afrasiabian), Shiraz, Iran, pp. 409-24.

Milanovic, P.T. (2000) *Geological Engineering in Karst: Dams, Reservoirs, Grouting, Groundwater Protection,Water Tapping, Tunnelling,* Zebra, Belgrade, 347 pp.

Milanovic, P.T. (2004) *Water Resources Engineering in Karst,* CRC Press, Boca Raton, 312 pp.

Palmer, A.N. (1991) Origin and morphology of limestone caves. Geological Society of America Bulletin, 103, 1–21.

Prohic, E., Peh, Z., Miko, S., 1998. Geochemical characterization of a karst polje. An example from Sinjsko Polje, Croatia. Environ. Geol. 33 (4), 263– 273.

Raeisi, E. (1999) Calculation method of karst water balance in Zagros Simple Folded zone, *Proceedings of the first Regional Conference on water balance,* Ahwaz, p. 39-49.

Raeisi, E. (2008) Ground-water storage calculation in karst aquifers with alluvium or no-flow boundaries, Journal of Cave and Karst Studies, v. 70, no. 1, p. 62–70.

Rahnemaaie, M. (1994) *Evaluation of infiltration and runoff in the karstified carbonatic rocks.* M. Sc. Thesis, Shiraz University.

Waltham T, Bell F, Culshaw M (2005) *Sinkholes and subsidence.* Springer, Chichester

Waltham, A.C. and P.G. Fookes (2005) Engineering classification of karst ground conditions , Speleogenesis and Evolution of Karst Aquifers, v. 3 (1), p. 1-19

Water Resources Investigation and Planning Bureau (1993). *Comprehensive study and research in water resources of the Maharlu karst basin (Fars).* Vol.1-4.

Williams, P.W. (1983) The role of subcutaneous zone in karst hydrology. J. Hydrol., 61: 45-67.

Hydrogeological Significance of Secondary Terrestrial Carbonate Deposition in Karst Environments

V.J. Banks[1,2] and P.F. Jones[2]
[1]British Geological Survey, Kingsley Dunham Centre, Nicker Hill, Keyworth, Nottingham
[2]University of Derby, Kedleston Road, Derby
UK

1. Introduction

A significant hydrogeological characteristic of karst environments is the precipitation of a proportion of the dissolved calcium carbonate derived from limestone dissolution. The study of such secondary deposits is important because they provide information on the palaeohydrogeology of the unsaturated zone at the time of precipitation. They also offer the potential to provide information with respect to climatic conditions through the study of stable isotopes and dating through the study of radiogenic isotopes. This chapter introduces the formational processes, depositional environments (hydrogeological, hydrogeochemical, biological and geomorphological) and post depositional history of secondary terrestrial carbonate deposits. Consideration is given to the associated research themes and techniques, in particular to the current research focus on the role of microbial communities in present day sediment-water interface processes (Pedley and Rogerson, 2010) and the implications for furthering the understanding of climate change and landscape evolution. These deposits have a world-wide distribution (Ford and Pedley, 1996; Viles and Goudie, 1990) and include speleothems, travertines, tufas, calcareous nodules, calcretes and carbonate cements, such that speleothems and tufa represent two end members of a continuum of freshwater carbonate (Pedley and Rogerson, 2010). They form in a range of climatic conditions, but are best developed in warm humid climates. Examples cited in the text include case studies from the White Peak, Derbyshire UK, which currently experiences a temperate humid climate and hosts a range of deposits as a consequence of its recent geological history. The White Peak was not subjected to glacial erosion during the most recent (Devensian, MIS 2-4) glaciation, therefore there is a potential for an extensive record of Quaternary palaeoclimatic conditions to be preserved in the secondary carbonate deposits.

2. Types of secondary carbonate deposit and their classification

Secondary terrestrial carbonate deposits include four major groups: *(i) Speleothems* that are characteristically deposited in caves above the water table (vadose or unsaturated zone) from saturated mineral solutions and can be seen at scales that range from millimetres to tens of metres. Typically, they are formed of calcium carbonate by the precipitation of calcite or aragonite from water as excess dissolved carbon dioxide is diffused into the cave atmosphere. Aragonite is a metastable polymorph of calcite, which predominantly occurs as acicular crystals in speleothems. Its occurrence is generally attributed to depletion of calcium ions in magnesium rich solutions. Other minerals, for example gypsum, can also be precipitated in this environment. They occur as *dripstones,* formed by water dripping from the ceilings or walls of caves (*stalagmites* or *stalactites*), or as *flowstones* formed on the walls or floors of caves. Speleothems can also form carbonate cements and cemented rudites in the cave environment. Typically they form as elongate (columnar) crystals perpendicular to the growth surface (palisade calcite, Kendall and Broughton, 1978), which may be visible as a series of growth layers. Speleothems commonly comprise alternations of soft and hard calcite (Ford and Williams, 2007) with most hard calcite occurring as palisade calcite, or as microcrystalline calcite. They also occur outside the cave environment, e.g. the anthropogenic flowstone precipitating from lime-rich cement at Lindisfarne Castle, Northumberland, UK (Figures 1 and 2). (ii)*Tufa* (ambient temperature, freshwater carbonate with carbon dioxide derived from the soil atmosphere) *and travertine* (carbonate precipitating from water that is hot as a consequence of deep circulation with carbon dioxide being derived from magmas and decarbonisation). These carbonates result from a combination of biologically moderated physicochemical processes and accumulate in a range of settings. They may take the form of cones, cements, barrages at metre to kilometre scale or minor, localised plant encrustations. Growth increments in tufas occur as contrasting laminae of dense micrite and more porous sparry crystalline calcite (Andrews and Brasier, 2005). (iii) *Carbonate cements* comprise minerals that fill pore spaces and bind particles together. Most terrigenous clastic sediments (rudites, arenites and argillites) have the potential to become cemented by minerals that fill the pore spaces. Although beyond the remit of this chapter, common cementing materials also include silica, iron oxide and sulphates. Carbonate cements occur externally to and within cave environments. Within caves they typically occur as cave breccias (deposits of calcium carbonate formed where cave water percolates into clastic sediments) and externally they typically comprise a greater variety of cemented rudites (e.g. fluvial gravels or screes; Figure 3). (iv) *Pedogenic carbonates* encompass caliche, soil nodules and rhizome (root) coatings, which may also exhibit incremental growth patterns. Fossilised forms include algal burrs, such as those of upper Jurassic age in Dorset, UK (Francis, 1984). These carbonates can be either inorganic or as consequence of biomineralization. Subsequent dissolution can lead to the development of third-order forms.

Secondary carbonate deposits attract a plethora of terminology (Ford and Pedley, 1996; Pentecost, 1995; Pentecost and Viles, 1994; Viles and Goudie, 1990, and Viles and Pentecost, 2007). Field descriptions are largely derived from the terms that are applied to microbialites (biolithite of Folk, or boundstone of Dunham [Tucker, 2011]), encompassing stromatolites and bioherms (Tucker, 2011; Viles and Goudie, 1990; see glossary). They may adopt columnar, planar, or oncolite forms. According to Pedley (1990) tufas can be described as either autochthonous (forming in-situ), e.g. phytoherm framestone (anchored) and phytoherm boundstone (dominated by the heads of skeletal stromatolites), or clastic (not anchored). Clastic tufas include: phytoclasts, oncoids, detritus, peloids and palaeosols. Petrological descriptions of cements are usually based on the extent of micrite (carbonate particles <4μm diameter) or sparite (clear, or white coarser equant calcite precipitated in pore space between grains; Tucker and Wright, 1990). Reference should be made to the literature for more unusual cement forms.

Fig. 1. Speleothem (flowstone) attributed to carbonate leaching from lime rich mortar, Lindisfarne Castle, Northumberland, UK.

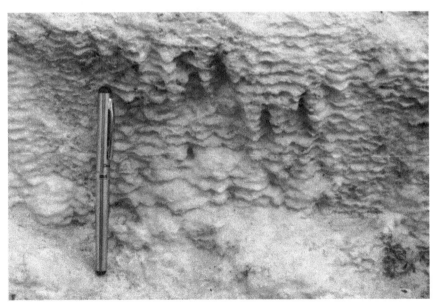

Fig. 2. Speleothem (flowstone) attributed to carbonate leaching from lime rich mortar, Lindisfarne Castle, Northumberland, UK.

Fig. 3. Cemented scree at Ecton, Manifold Valley, Staffordshire, U.K.

Classification and interpretation of depositional environments is fundamental to the applied geological aspects of secondary carbonate deposits. Owing to the breadth of depositional environments and their global distribution, a number of potentially useful classification schemes have been developed. Classifications that have been applied to the major groups follow.

2.1 Speleothems

Speleothems are generally classified according to their morphology or their origin (Hill and Forti, 1995). The former is used more frequently and Hill and Forti (1995) use the term *speleothem type morphology* to define a distinct morphology that is controlled by one or more hydrological mechanisms, e.g. dripping, flowing, pool, geyser, capillary, condensation and aerosol water. More broadly, speleothems can be classified as gravitational (dripstones and flowstones), which are the key focus for this chapter, or as non-gravitational erratic forms, including coverings or shields, helictites (capillary-fed), botryoidal forms, moonmilk, pendants, straws, cave pearls, rimstones and pool deposits. Each of these is associated with a specific setting within the cave environment (Ford and Williams, 2007; Hill and Forti, 1995). Whilst stalactites and stalagmites are formed by dripping water, flowstone is formed layer upon layer by water flowing over surfaces. Draperies (e.g. curtains) may be formed by both processes separately or in combination. As a consequence of their layered structure, gravitational speleothems can be dated and contribute evidence of local and regional tectonic histories. They can also be used in conjunction with stable isotope analyses for the interpretation of climate change (section 6). Where flowstones occur, indicators of historic flooding may be preserved on cave walls as erosional features or inclusions of detritus.

2.2 Travertine and tufa deposits

The basis for the classification of the depositional environments associated with travertines and tufas comes from the work of Chafetz and Folk (1984) who defined five main classes: waterfalls; lake-fill; sloping mound or fan; terraced mound and fissure ridge. A number of classification systems have followed, which are broadly based on depositional setting, geomorphology or biology (Pazdur et al., 1988; Pedley, 1990; Pentecost and Lord, 1988; Pentecost and Viles, 1994; Viles and Pentecost, 2007). Additionally, classifications have been devised for engineering purposes; such classifications need to take account of the heterogeneous nature of tufa, e.g. the engineering classification of tufa in the Antalya area, SW Turkey (Dipova, 2011), which was derived from a consideration of primary fabric, diagenesis and strength. Pedley (1990) subdivided tufas on the basis of their environmental setting (Table 1) and this provides the reference point for subsequent descriptions in this chapter.

Some tufa deposits occur where deep hydrothermal flow resurges, commonly in association with faults. Due to the absence of associated assemblages of deposits, these tufas would likely be classified in the perched spring-line model. Examples include the deposits at Matlock Bath, Derbyshire, UK (Pentecost, 1999). As with speleothems, growth couplets have been identified as representing annual seasonality (Andrews and Brasier, 2005), which offer the potential for palaeoclimatic and palaeoenvironmental interpretation (section 6). An unusual non-biogenic occurrence of tufa occurs in ultramafic rocks in northern Oman, where hyperalkaline groundwaters precipitate tufa (Clarke and Fontes, 1990).

Pentecost and Viles (1994) classify tufa as a form of travertine and they distinguish between meteogene and thermogene travertine on the basis of the source of the carbon dioxide (soil and deep crustal sources respectively). Thermogene travertine is associated with volcanic centres, high carbon dioxide discharges and high geothermal gradients (section 3.5). Pentecost and Viles (1994) presented a further classification for thermogene travertines, subdividing: spring (fissure ridge or mound, including those of saline lakes) river (cascade, cemented rudites and barrages) and lake deposits (crusts).

Model	Deposits
Perched springline. (resurgences part way up slopes).	Proximal: dominated by liverworts and bryophytes; colonisation by cyanophytes and diatoms. Distal: Fine intraclast tufa and microdetrital tufa.
Cascade.	Curtains of moss associated with waterfalls. Notable absence of upstream lake sediment and biotal associations.
Braided fluviatile.	Braided cyanolith-dominated deposits comprising oncoids and micro-detritus.
Fluviatile barrage.	Phytoherms that obstruct stream flow forming barrages with associated upstream lake sediment and biotal associations.
Lacustrine.	Macro- and microphytes that characterise lake margins; stromatolite, oncoid and intraclast tufas characterise the shallow water and micro-detritus the marginally deeper water.
Paludal.	Surface coatings of tufa on vegetation in marshy localities, where resurgences occur on poorly drained slopes or alluvial valley bottoms.

Table 1. Classification of tufa based on environmental setting (Pedley, 1990).

An outcome of an investigation of calcretes and speleothems in deep time (Brasier, 2011) was the observation that the relative absence of biogenic soils prior to the evolution of vascular plants implies that different processes were associated with the deposition of terrestrial carbonates during the Archaean, Proterozoic, Cambrian, Ordovician and Silurian. Accordingly, Brasier (2011) has suggested that the term tufa cannot really be applied to deep time; instead, more descriptive classifications, e.g. spring carbonate, stream carbonate and lacustrine carbonate may be more appropriate.

2.3 Carbonate cements

A range of clastic sediments can be cemented, but coarser sediments contain larger pore spaces, which allow thicker cement rinds to form and render them more favourable for research purposes. Detailed petrological descriptions facilitate classification on the basis of vadose or phreatic cementation. A good case study by Strong et al. (1992) described the range of vadose fabrics associated with the cementing of glacial gravels in North Yorkshire, England, UK. This case study also includes a description of some of the terms used to describe pedogenic carbonates, including rhizocretions (the tubiform cements that form around plant roots). Additional pedogenic forms (Tucker, 2011) include duricrusts (lithified, pedogenic surface layers), laminated layers and vadoids (laminated spherical grains that are commonly biologically mediated).

2.4 Pedogenic carbonate

These carbonates primarily occur in arid, semi-arid, or subhumic climates and can be classified on the basis of either morphology or formational process. Irrespective of climatic conditions, formational processes can be subdivided into: per descensum, per ascensum, in situ and biogenic models (Curtis, 2002).

2.5 Third order deposits

Regional lowering of groundwater levels can induce dissolution of secondary carbonate cements. Given their topographical setting, cascade tufa deposits are particularly vulnerable to this form of weathering. However, active dissolution can also result in saturation and supersaturation of the infiltrating water with a potential for re-precipitation, particularly in association with biological mediation. Typical of this process are the third order speleothem deposits observed in the Via Gellia, near Cromford, Derbyshire, UK, which take the form of micro-stalactites (Figure 4) and flowstone. Similarly, inundation of calcareous aeolianites is often characterised by vertical piping because dissolution occurs where percolation is guided by tree roots (Ford and Williams, 2007).

Fig. 4. Micro-scale third order speleothems in tufa in the Via Gellia, near Cromford, Derbyshire, UK. Larger examples also occur. Binocular microscope image: the British Geological Survey Mineralogy and Petrology Laboratories.

3. Depositional processes associated with secondary carbonate precipitation

Secondary carbonates result from a carbonate source-pathway-receptor system that operates at or very close to the Earth's surface. This generally involves dissolution of an existing deposit of calcium carbonate followed by transport of the dissolved species via surface and groundwater flow paths to a point of precipitation. Favourable conditions for precipitation include supersaturation with respect to calcium carbonate and sites where the physical perturbation of water promotes carbon dioxide outgassing.

3.1 Hydrogeochemistry

The hydrogeochemistry of secondary carbonate deposits is fundamental to understanding their formation. The following comprises a brief overview of the carbonate system. The interested reader is recommended to refer to additional texts, including: Ford and Williams (2007) and Kehew, (2001), as well as the other references cited.

Limestone dissolution has been studied extensively in the context of karst geomorphology (Appelo and Postma, 2005; Ford and Williams, 1989 and 2007; Gunn, 1986). The controlling chemical equations with respect to the dissolution of calcium carbonate are:

$$CO_2(g) \leftrightarrow CO_2(aq) \tag{1}$$

This equation represents the absorption of carbon dioxide in water. At equilibrium, the activity of dissolved carbon dioxide is proportional to the partial pressure of carbon dioxide in the gas phase in contact with the aqueous phase containing the dissolved carbon dioxide. Atmospheric P_{CO_2} is $10^{-3.5}$ atm., whereas the P_{CO_2} values for groundwater are typically an order of magnitude higher (up to approximately 6%, Ford and Williams, 2007), as a consequence of the addition of biologically (plant respiration) derived carbon dioxide (Atkinson, 1977). The variation in the production of carbon dioxide is primarily related to the temperature, moisture content and amount of organic matter in the soil and therefore it reflects climate and seasonality (Ford and Williams, 2007; Kehew, 2001).

$$CO_2(aq) + H_2O \leftrightarrow H_2CO_3 \tag{2}$$

Aqueous carbon dioxide dissolves in water to form carbonic acid, which is a weak acid with a potential for dissociation. The theoretical assumption is that all of the CO_2 occurs as carbonic acid, whereas in practice, most of the CO_2 is present as dissolved CO_2. The reaction between aqueous CO_2 and water is slow compared to reactions involving H_2CO_3, thus it becomes rate limiting (Kaufmann and Dreybrodt, 2007). Carbonic acid is a diprotic acid, i.e. it can dissociate twice:

$$H_2CO_3 \leftrightarrow H^+ + HCO_3^- \tag{3}$$

$$HCO_3^- \leftrightarrow H^+ + CO_3^{2-} \tag{4}$$

The three inorganic carbon species that result from the dissolution of carbon dioxide in water are active over differing pH ranges: H_2CO_3 predominates in acid conditions (pH 1 to 6.4); at pH 6.4 the activities of HCO_3^- and H_2CO_3 are equal; from pH 6.4 to 10.33 HCO_3^- is more active; at pH 10.33 the activities of HCO_3^- and CO_3^{2-} are equal, with CO_3^{2-} being most active where the pH exceeds 10.33.

Additional hydrochemical factors to be considered in the context of limestone dissolution include: (i) The common ion effect, whereby a common ion derived from the dissolution of a more soluble mineral will reduce the solubility of the less soluble mineral, e.g. the addition of sodium bicarbonate to a solution in equilibrium with calcite would cause an increase in the saturation index for calcite, thereby resulting in calcite precipitation. (ii) Incongruent dissolution, which can occur because of the differing activities of minerals at different temperatures or as a consequence of differing reaction rates. An example of the former is the incongruent dissolution of dolomite and simultaneous precipitation of calcite as a consequence of the common ion effect imparted by anhydrite, which results in dedolomitization (Bischoff et al., 1994). (iii) Increases in ionic strength, which cause reductions in activity coefficients and consequential increases in solubility.

Dissolution in karst environments (karstification) can result in surface lowering or an increase in the underground permeability. Rates of dissolution can be calculated from the product of the discharge volume and solute (calcium carbonate) concentration. Both ground surface and underground dissolution are not uniform processes. Dissolution is greatest: at points of convergence where mixing corrosion can occur; in zones of more intense biological activity; where soil moisture is high, and in areas with a sunny aspect (Ford and Williams, 2007). As a consequence, dolines can become the focal point for higher rates of surface dissolution. In the UK, dissolution rates of 83 m^3 km^2 a^{-1} have been quoted for Derbyshire and Yorkshire (with 0% and 50% respectively being derived from underground sources; Pitty, 1966 and Sweeting, 1966). A rate of 69 m^3 km^2 a^{-1} was measured by Gunn (1981) for Waitomo, New Zealand (37% from the soil profile and the remainder from up to 5- 10 m of bedrock) where the epikarst is better developed than Derbyshire and Yorkshire. Any assumption of 100% surface lowering is likely to be in error. More recent developments in the understanding of carbonate dissolution, in particular that of kinetic thresholds (see below) and the importance of flow convergence, suggest that an underground component should be expected in all environments. Ford and Williams (2007, p82) note that *"hundreds of studies of solutional denudation have been completed since 1960. A major shortcoming of much of the work is that autogenic rates have often not been distinguished from mixed autogenic-allogenic rates so that there is still no unequivocal answer to the question posed long ago by climatic geomorphologists: in which climatic zone does karst evolve most rapidly?"*

Initial openings in unconfined karst settings are slowly enlarged by groundwater that is close to saturation with respect to calcium carbonate. The routes conducting the highest discharge are subject to greater dissolution (Palmer, 1991, 2002). Consequently, conduit development is enhanced by larger initial openings and increased hydraulic gradients such that conduit initiation is commonly attributed to base level lowering, which may be a consequence of uplift. For a given partial pressure of carbon dioxide, the initial rate of dissolution decreases in an approximately linear manner with increasing calcium carbonate content, but at 60-90% saturation the dissolution rate decreases rapidly (the kinetic threshold). Whilst the slow uniform dissolution that delays the final stage of saturation with respect to calcite facilitates the gestation of long conduit flow paths and enables deeper penetration of nearly saturated water, it may inhibit supersaturation with respect to calcite. Whereas flow in pre-dissolutional openings is laminar, as dissolution proceeds and openings are enlarged to the hydrodynamic threshold (generally considered to be 10 mm, Fetter, 2001), turbulent flow develops and the rate of carbonate rock dissolution increases rapidly as circulating water becomes less saturated. The hydrodynamic threshold also coincides with the kinetic threshold and the onset of clastic sediment transport, which further contributes to dissolutional enlargement (White, 2002). The rate of dissolution increases with discharge until a maximum is achieved. Flooding increases flow rates and flood water is generally more aggressive being characterised by a lower pH and lower concentration of calcium. During flooding, additional flow paths will be activated, both in the vadose and the phreatic zones, facilitating rapid dissolution and increasing the efficiency of the system.

Even without topographical and hydrological focusing of flow, limestone dissolution would be non-uniform because some limestones have a greater propensity for dissolution than others. For example, dissolutional activity (speleogenesis) tends to focus on inception horizons (Lowe, 2000; Banks et al., 2009). Typically, inception horizon-guided dissolution focuses on incipient physical, lithological, or chemical differences that form a focal point for conduit gestation. These differences include: breaks in sedimentary style or bedding; variation in trace chemistry; fossil bands, or contrasts in permeability (e.g. boundaries between limestone and chert, shales or clays). Given sufficient residence time, rock-groundwater interaction ensures that the groundwater chemistry reflects the chemistry of the host geology. Thus inception horizons may also form a source of calcium carbonate, despite the fact that the rates of dissolution quoted above suggest that conduit sources generally provide less than 50% of the dissolved calcium carbonate in a catchment.

Other sources of calcium carbonate include soils, superficial deposits and a range of anthropogenic materials. Soil and superficial deposits derived from limestone terrains can contain a significant proportion of leachable calcium carbonate. Typical artificial sources of carbonate with a greater propensity for dissolution include former waste tips resulting from lime processing, quarrying or mining. For example, the source of the artificial tufa barrages in Brook Bottom (Figure 5), which are in the order of 1 m high (Ford and Pedley, 1996), and the source for a nearby, rapidly accumulating speleothem in Poole's Cavern, near Buxton, Derbyshire, UK   (Baker et al., 1999) comprises waste heaps related to the former lime manufacturing industry (Figure 6). The formation of pedogenic carbonate with carbonate derived from artificial sources such as slag, construction materials and cement has been described by Manning (2008).

Locality	Secondary carbonate deposit type	Rate (mm/annum)	Reference and notes
Caves of Niue Island, South Pacific.	Speleothem	0.23-0.34	Aharon et al., 2006.
Lathkill Dale, Peak District, Derbyshire, UK.	Tufa	2.5	Andrews et al., 1994. Holocene accumulation rate.
Poole's Cavern, Peak District, Derbyshire, UK.	Speleothem	2.1 – 5.0	Baker et al., 1999. Associated with former lime kilns. 1910 – 1996.
Palaeozoic caves of Wallonie, Belgium.	Speleothem	0.5 – 2.17	Genty and Quinif, 1996.
Malham Tarn, North Yorkshire, UK.	Tufa	0.01-1.30	Pentecost, 1978.
Goredale Beck, North Yorkshire, UK.	Tufa	1-8	Pentecost, 1978.
Mato Grosso, Brazil.	Speleothem	0.012	Soubiès et al., 2005. Variable growth rate.

Table 2. Examples of secondary terrestrial carbonate accumulation rates.

The precipitation of calcium carbonate can be summarised by:

$$Ca^{2+} + 2\,HCO_3^- \leftrightarrow CaCO_3 + CO_2 + H_2O \tag{5}$$

A state of supersaturation is required for carbonate precipitation. This can be brought about by the degassing of calcium carbonate enriched waters (Chen et al., 2004; Lorah and Hermon, 1988) that may be associated with cooling and physical or biological degassing of carbon dioxide, reducing the amount of calcium carbonate that can be held in solution. A further process leading to supersaturation is that of evaporation, which is considered to be

Fig. 5. Anthropogenic tufa barrages, Brook Bottom, Derbyshire, UK.

Fig. 6. Cemented lime kiln waste, Brook Bottom, Derbyshire, UK.

the cause of tufa precipitation near many cave and mine entrances (Ford and Williams, 2007). However, a solution that is supersaturated with calcium carbonate does not necessarily give rise to carbonate precipitation if appropriate nucleation sites are not present. Rates of carbon dioxide out-gassing can exceed rates of calcium carbonate precipitation resulting in supersaturation, with a potential to mask any groundwater mixing effects (Thrailkill, 1968). In the case of speleothems, where there is a lower partial pressure of carbon dioxide in the cave atmosphere than in the incoming water, degassing of carbon dioxide leads to supersaturation and consequential precipitation of calcium carbonate.

Accumulation rates for secondary terrestrial carbonates can provide valuable information with respect to landscape evolution. The relatively rapid accumulation rates are such that measurements can be determined in mm per year; examples are presented in Table 2.

3.2 Biological mediation

The nature and extent of biological mediation (enabling) of secondary carbonate precipitation reflects the physical setting of the deposit. Various microbes, flora and fauna contribute to tufa deposition. Cyanobacteria are usually the dominant microbial component (their calcification being associated with the mucopolysaccharide) in fast-flowing streams supersaturated with respect to calcite, where sheath encrustation is the dominant form of calcification (Riding, 2000). Cyanobacteria mineralization is extensively associated with tufa precipitation (Andrews and Brasier, 2005; Brasier et al., 2011; Pentecost, 1988). Photosynthesis causes alkalinization, while exopolymeric substance (EPS) acts as binding site for calcium (less so in freshwater tufas; Dittrich and Sibler, 2010) and the consequential focus for calcium carbonate precipitation. Similarly, microbes are also present in many cave systems, varying between the twilight and aphotic zones of caves (Jones, 2010). They include a diverse range of algae, actinomycetes, bacteria, fungal hyphea and cyanohycea. Cave microbes contribute to the destructive (dissolutional substrate breakdown, boring and degradation) and constructive processes (trapping and binding of detrital particles, calcification and precipitation) that influence the growth of speleothems (Jones, 2010). Evidence of these processes comes from the presence of fabrics (e.g. microbial stromatolites), mineralized microbes and geochemical markers (e.g. lipid biomarkers; Jones, 2010). Cave microbial processes reflect the habitat, particularly the light distribution.

When microbes are encrusted and replaced by calcite, they become part of the substrate. This process results in a range of calcite crystal forms, in both tufas and speleothems. Biomineralization is associated with: bacterial cells, including picocyanobacteria (unicellular cyanobacteria with a cell diameter of 0.2 to 2.0 μm); sheaths, and EPS. Biogenic mineralization can occur through either biologically controlled or biologically induced processes (González-Muñoz et al., 2010). Biologically controlled mineralization occurs in isolated compartments within a living organism, resulting in highly ordered mineral structures (González-Muñoz et al., 2010), which are more typical of shells, but uncommon in bacteria. Biologically induced mineralization is the result of microbial metabolism. There are two stages involved: firstly, active modification of the physical chemistry in the environment of the bacteria leading to an increase in ion concentration (supersaturation; equations 6 to 7); secondly, nucleation of mineral (equation 7). Homogeneous nucleation requires a higher degree of supersaturation, whereas heterogeneous mineralization results from nucleation on bacterial cell walls, bacterial EPS or the new mineral phase (González-

Muňoz et al., 2010). The geochemical equations involved in bacterial ion concentration (González-Muňoz et al., 2010) can be summarised as:

$$HCO_3^- + H_2O \rightarrow (CH_2O) + O_2 + OH^- \tag{6}$$

This equation represents the photosynthetic bacterial conversion of bicarbonate into reduced carbon. Similar effects can be produced by bacteria that produce ammonia by oxidative deamination of amino acids. Where carbon dioxide is generated by the bacterium, supersaturation with respect to bicarbonate or carbonate may result (González-Muňoz et al., 2010).

$$HCO_3^- + OH^- \rightarrow CO_3^{2-} + H_2O \tag{7}$$

Equation 7 results from the exchange of intracellular hydroxide ions for extra cellular bicarbonate ions across the cell membrane. Alkalinization around the bacterial cells induces carbonate generation.

$$Ca^{2+} + CO_3^{2-} \rightarrow CaCO_3 \tag{8}$$

Equation 8 represents the precipitation of calcium carbonate on the cell surface. Calcium carbonate may first be precipitated as vaterite. Evidence for a two phase process in active tufa deposition in the Via Gellia, Derbyshire, UK, can be seen in Figures 7 to 9. However, different cyanobacterial species exhibit different calcification fabrics (Pentecost, 1991).

Fig. 7. Active tufa precipitation in the Via Gellia, Near Cromford, Derbyshire, UK. Binocular mircroscope images: the British Geological Survey Mineralogy and Petrology Laboratories.

Dittrich and Sibler (2010) analysed and modelled the functional groups of extracellular polysaccharides of three picocyanobacteria establishing the presence of five to six surface sites, corresponding to: carboxyl, phosphoric, sulphydryl, amine phenol, and hydroxyl groups. The carboxyl and carboxyl-phosphoric groups dominated in all strains, closely followed by the hydroxyl groups. Polysaccharides were found to be negatively charged at a

Fig. 8. Active tufa precipitation in the Via Gellia, Near Cromford, Derbyshire, UK. Binocular microscope images: the British Geological Survey Mineralogy and Petrology Laboratories.

Fig. 9. Active tufa precipitation in the Via Gellia, Near Cromford, Derbyshire, UK. Binocular microscope images: the British Geological Survey Mineralogy and Petrology Laboratories.

pH range of 6 – 7. Therefore, calcium ions can easily be attracted to them. However, removal of calcium reduces the degree of saturation, thereby inhibiting calcium carbonate precipitation. The presence of the carboxyl groups offers the potential to remove metals, thereby overcoming the inhibition to calcium solubility resulting from the presence of low concentrations of metals (Terjesen et al., 1961; Dittrich and Sibler, 2010). Decomposition of the EPS releases bicarbonate and calcium ions, which increases the calcium carbonate saturation state and promotes precipitation.

3.3 Geomorphological and tectonic mediation

A number of deposits are representative of specific geomorphological and tectonic settings, as acknowledged in Pedley's (1990) classification of tufas (section 2.2) and as evident in the

Fig. 10. Pamukkale Travertine, SW Turkey. Photograph: Anthony H Cooper, British Geological Survey.

Antalya area of south-west Turkey, where tufa terraces have been related to glacio-eustatic sea level change (Glover and Robertson, 2003). Similarly, Forbes et al. (2010) described a number of tufa deposits in the south-western coastal zone of Western Australia. These deposits are characterised by cascade to barrage pool and perched spring line to barrage pool situations, which are associated with coastal waterfall and supratidal geomorphological settings respectively. Geomorphology and tectonic setting can also influence the carbonate source and flow paths through uplift and erosion. As an example, a rock slide in the Fern Pass, Austria produced carbonate rock flour which formed the carbonate source for subsequent cementation of the rockslide breccias (Ostermann et al., 2007). As thermal deposits, travertines are also likely to be moderated by tectonic events, particularly given that the location of these deposits is commonly related to faults (Pentecost, 1995). Similarly, fissure ridge travertine deposits, associated with listric faulting in the Gediz Graben extensional province of Turkey have been explained by their tectonic setting (Selim and Yanik, 2009). The Pamukkale travertine (Figure 10) in Turkey occurs in a different extensional tectonic regime (Selim and Yanik, 2009; Şimşek, 1993). In the case of the Lapis Tiburtinus travertine, Central Italy, travertine cycles reflect water table fluctuations associated with fault and volcanic activity between 115 000 and 30 000 BP (Faccenna, et al., 2008).

3.4 Anthropogenic influences

As well as providing artificial sources of calcium carbonate, anthropogenic influences can affect flow paths and the depositional rates of secondary carbonates that are commonly encountered in infrastructure, mines and industrial areas. The growth of speleothems has been reported in a number of such settings, e.g. railway arches and disused water reservoirs. Recent work in opening the disused railway cuttings between Buxton and Bakewell in the White Peak, Derbyshire, UK, has exposed occurrences of speleothems, flowstone and tufa associated with the engineered structures, including tunnels, bridges and cuttings. Occurrences of this kind may be a consequence of either the opening of new flow paths, or the interception of pre-existing ones, as observed in a cutting (Figure 11) where tufa precipitation is clearly associated with water discharging from exposed inception horizons. Flow rates associated with these deposits are generally low, with flow commonly occurring as minor seepages in the unsaturated zone. The nature of anthropogenically mediated secondary deposits would appear to reflect the significance of biological mediation, with tufa forming in moist locations where mosses and bryophytes establish themselves, commonly associated with shade and good ventilation. As the majority of the engineered structures can be dated, secondary carbonate deposits in these settings provide an opportunity for assessing minimum rates of precipitation. Cemented kiln waste encountered in Brook Bottom, Derbyshire, U.K (Figure 6), provides further evidence for the occurrence of anthropogenically moderated calcite precipitation. Underground, rapid rates of calcium carbonate precipitation are implicit in the occurrence of tufa deposits above a skeleton discovered in historic mine workings in Lathkill Dale, Derbyshire, UK in 1744 (Rieuwerts, 2000).

Anthropogenic influences have been linked with a range of factors that may have contributed to the post-Holocene decline in tufa deposition (Goudie et al., 1993), which in Derbyshire, UK, primarily occurred approximately 4000 years BP. This coincides with Neolithic to Early Bronze Age deforestation (Taylor et al., 1994), which may have influenced barrage type (section 2.2) tufa deposition in one or more of the following ways (Goudie et al., 1993): reduced availability of carbon dioxide in the vicinity of areas of groundwater recharge; less woody material available to form dams; induced soil erosion, associated with further reductions in the availability of carbon dioxide; alteration to the shade causing stress to mosses or cyanobacteria at resurgences; increased access for grazing with consequential changes to groundwater chemistry, including eutrophication, and turbidity, thereby inhibiting the bacteria and mosses responsible for tufa accumulation, and increased surface run-off causing increased turbidity and dissolution potential during periods of high discharge. Tufa formation is severely limited when the mean annual air temperature falls below 5^0C (Pentecost, 1996; Pentecost and Lord, 1988), which suggests that a component of the late Holocene decline may be associated with lower mean air temperatures.

Fig. 11. Tufa formation below an inception horizon in a railway cutting between Buxton and Bakewell, Derbyshire, UK. 1 m length of tape measure for scale.

3.5 Travertine forming processes

Pentecost (1995) summarised the distribution of Quaternary thermogene travertine formations in Europe and Asia Minor. Of the 93 sites identified, 56 are known to be active and many of the European deposits date to the Pleistocene or late Pliocene. A significant proportion of the active sites occur in Italy and Turkey, where they correspond with volcanic centres and high carbon dioxide discharges, associated with high geothermal gradients. The precise relationship between volcanism and travertine formation has yet to be established. On the basis of isotope analyses, it has been hypothesised that carbon dioxide enriched fluids derived from the upper mantle and from limestone decarbonation dissolve sedimentary carbonate (Pentecost, 1995). However, in the case of the Lapis Tiburtinus travertine, Tivoli, Central Italy, carbon isotope analysis of the travertine suggests a hydrothermal origin for the fluids that precipitated the tufa (Faccenna et al., 2008). The majority of hydrothermal springs rise on faults (Pentecost, 1995). The largest depositional region extends from Greece, through Turkey and across the Caucasus Mountains into Russia, where convergence of the Eurasian and Afro-Arabian plates is believed to be responsible for the recent volcanism and hot spring activity.

3.6 Diagenesis

Diagenesis of carbonate sediments can occur through cementation, dissolution, microbial micritization or neomorphism. The metastable condition of many carbonate minerals, such as aragonite, renders them susceptible to recrystallization. Potentially, this may result in changes that cause the deposits to be less useful for interpreting climate change or depositional environments. Recrystallisation may occur through changes to the isotopic signature or fractionation of elements used for dating. Thus, understanding of diagenetic processes is important. Martín–García et al. (2009) demonstrated this in the case of speleothems from the Castañar Cave, in the southern part of the Iberian Massif in Spain. Here speleothems in dolomite hosted karst were found to have undergone micritization and neomorphism thereby modifying the primary features including the stable isotope, strontium and magnesium contents. Similarly, in a study of the diagenesis of aragonite in speleothems in Korea, Woo and Choi (2006) found that aragonite inversion to low magnesium calcite was associated with the remobilisation of stable isotopes and trace elements, with a notable difference in the carbon isotope signature.

In contrast, in a petrographic examination of Greek tufa deposits with a significant development of sparite over micrite, the relative absence of diagenetic processes was established (Brasier et al., 2011). This disproved a formerly held view that the extent of sparite was likely to be a secondary feature (diagenetic aggrading neomorphism, e.g. Love and Chafetz, 1988) of the tufa. The study of the distribution of sparite and micrite in two Greek Pleistocene tufa stromatolites (Brasier, et al., 2011) revealed primary columnar calcite spar in a younger deposit (ca 100 ka) from Zemeno occurring immediately above chironomid larval tubes with which its growth was associated, whilst an older tufa (ca 1 Ma) from Nemea comprised proportionately more sparite with some chironomid tubes and cyanobacterial filaments. Comparison of stable isotopic trends (section 5.2) revealed that both deposits supported resolvable seasonal responses, suggesting that there had been limited post-depositional alteration of either tufa. The higher proportion of sparite in the Nemea deposit has been attributed to abiotic, speleothem-like growth of near hemispherical laminations from thin films of water (Brasier et al., 2011).

4. Spring and cave drip geochemistry

The use of seasonal signatures derived from carbonate deposits, particularly as a source of high-resolution palaeoclimate data, drives the need to understand how this is expressed in the supply waters. Spring geochemistry is of interest to the research of secondary terrestrial carbonate deposits, because it reflects the source of the carbonate as well as the processes operating during transport of the solutes between the points of dissolution and precipitation. Drip and cave air geochemistry is of equivalent interest in the case of cave deposits (Baldini et al., 2006). Calibration of speleothem oxygen isotopes in current calcite deposits at Tartair cave, Sutherland, North West Scotland with the oxygen isotope signatures of precipitation and percolating waters (Fuller et al., 2008) demonstrates the benefit of speleothem research to climate studies.

4.1 Flow path geochemistry

Groundwater can be characterized by the degree of equilibrium between the water and the wall rock (Drake and Harmon, 1973; Richardson, 1968; Shuster and White, 1971, Smith and Atkinson, 1976). Factors that influence flow path geochemistry include: flow-through time (water-soil contact or water-rock contact time); atmospheric conditions, including temperature; thickness and type of superficial cover; bedrock geology, and epikarst thickness. These variables are reflected in the range of proposed classifications, based on: individual parameters, ratios of parameters (Downing, 1967 and Vervier, 1990), flow-through times, as indicated by hardness and Pco_2 (Drake and Harmon, 1973; Pitty, 1966), seasonal variation of parameters (Shuster and White, 1971) and variation with discharge (Jacobson and Langmuir, 1974). Kehew (2001, p. 16) suggested, *"By knowing the state of equilibrium between the water and minerals within the aquifer, we can predict the type of reactions that are occurring or would be likely to occur"*, but it is rarely possible to sample a karst system along its flow path, instead use is made of springs and geochemical modelling to assess the likely processes within the system.

Conceptually, speleothems form within a flow path, thereby providing information on flow path geochemistry in the vadose zone. It has been established that the five main controls on their growth rates are: drip rate, activity of calcium in the drip water, air temperature, cave air Pco_2 and film thickness (Fairchild et al., 2001). In this context, cave drip water chemistry and cave air chemistry is important for understanding the hydrological controls on speleothem growth rates. This has been demonstrated in studies of hourly resolved cave Pco_2, and cave drip water hydrochemical data, in Crag Cave, SW Ireland (Baldini, et al., 2006; Sherwin and Baldini, 2011; Tooth and Fairchild, 2003). These studies demonstrated that calcite deposition on stalagmites can be moderated by prior calcite precipitation on short timescales. Given relatively constant conditions (air temperature, film thickness and low drip rates) controls on speleothem growth were most strongly influenced by cave air in winter, whilst drip water dilution caused by rain events may play a more significant role during the summer. In a separate study of the same cave system, Tooth and Fairchild (2003) established that some speleothems at this site may record a signal of palaeohydrology determined by variations in Mg/Ca ratios, with higher Mg/Ca ratios indicating lower flow conditions when base flow is maintained by long residence time storage water.

4.2 Spring geochemistry

Geochemistry can be used to assess the functioning of karst springs, for instance the degree of equilibrium between the groundwater and the wall rock has been taken as an indicator of the type of flow feeding the spring (e.g. Atkinson, 1977a; Shuster and White, 1971; Worthington and Ford, 1995). Shuster and White (1971) classified diffuse and conduit flow waters; Bertenshaw (1981) vadose and phreatic, or open and closed systems, and Worthington (1991) underflow and overflow systems. The terms "open" and "closed" are defined by Appelo and Postma (2005), Ford and Williams (2007), Gunn (1986) and Smith and Atkinson (1976). The open system is one in which gas, water and rock are all in contact with one another such that carbon dioxide is available to replace that used up in the reaction of limestone and carbonic acid. The closed system is one in which gas and water come into equilibrium, but a replacement supply of carbon dioxide is not continuously available during the reaction between limestone and carbonic acid. There is a gradation between fully open and fully closed systems. In the closed system, the concentration of carbonate species with changing pH is non-linear and less than in open systems. Karst systems are also described as diffuse or focused, which again may be reflected in the spring chemistry. Further understanding of aquifer processes can be derived from the seasonality of spring data.

4.3 Seasonality in spring and groundwater geochemistry

Seasonality in biogenic activity results in fluctuation in the biogenic production of carbon dioxide with a consequential seasonality to the partial pressure of carbon dioxide in infiltrating groundwater. Additionally, maximum evapotranspiration associated with the summer months concentrates constituents entering the ground to as much as twice that measured in the atmospheric precipitation (Edmunds, 1971). The extent to which seasonality is exhibited in spring water chemistry has been the subject of debate, possibly in part due to the method of monitoring (analogue versus digital). Shuster and White (1971) were able to use seasonality as an indicator of flow type in the Central Appalachians. More specifically, they related the variability of total hardness (expressed as a percentage coefficient of variation i.e. standard deviation/mean) to the type of resurgence. Springs with a variability of greater than 10% were interpreted as conduit flow, whilst those with less than 5% as diffuse, or percolation flow. However, in the Central Appalachians the variation in carbon dioxide pressures was more closely related to source areas (Shuster and White, 1971) than to seasonality. Deriving comparable coefficients of variation with respect to total hardness, Jacobson and Langmuir (1974) concluded that discharge was a more important influence on water chemistry than season, particularly for dispersed recharge type springs. In a more recent study in subtropical areas of SW China, Liu et al. (2007) collected data over two years of continuous monitoring of pH, conductivity, temperature and water stage from two epikarst springs (Nongla and Maolan springs). This was used to calculate partial pressures of carbon dioxide and the saturation indices with respect to calcite and dolomite. The study identified marked seasonal, diurnal and storm-related variations in the monitored and modelled parameters. Coefficients of variation of the parameters indicated that the greatest variation was at the seasonal scale, whilst storm scale exceeded diurnal variation. The variation was marked by higher conductivity and lower pH in the summer and daytime. Co-variation with temperature indicates that this influences the production of carbon dioxide in the soil. High rainfall events mask the seasonality as a consequence of the dilution. Monthly monitoring, over a 5-year period at fourteen localities along a tufa bearing stream on Carboniferous to Permian limestone at Shimokuraida, Niimi City, in southwest Japan, identified seasonal variation in the soil and spring water partial pressures of carbon dioxide. The variation was higher in the summer to autumn and lower in the winter to spring (Kawai et al., 2006).

4.4 Relationships between supply water chemistry and carbonate precipitation

Individual studies have demonstrated how supply water chemistry influences carbonate precipitate geochemistry. For example, the oxygen isotope content of speleothems in Tartair

Cave, Sutherland, North West Scotland corresponds with the oxygen isotope signatures of precipitation and percolating waters (Fuller et al., 2008; section 4). Leybourne et al. (2009) demonstrated that rapidly deposited tufa Sr/Ca ratios were controlled by spring water Sr/Ca ratios in a carbonate rock aquifer in the Interlake Region of Manitoba, Canada. They showed that whilst the $\delta^{18}O$ in the tufa is in equilibrium with the water, the $\delta^{13}C$ is enriched compared with the groundwater. In a case study in the Wye catchment, Derbyshire, UK, Banks et al. (2009) defined a number of formationally-based hydrogeological domains. These were derived from the results of dye tracing tests, analysis of water well monitoring and spring geochemistry and established that occurrences of barrage tufa are associated with only one of the domains. The implication was that the inception horizons in this unit provided a source of calcium during tufa precipitation (Banks et al., 2011).

4.5 Stream chemistry

Whilst spring chemistry provides an indication of the functioning of the karst aquifer, stream or river chemistry in karst terrains provides a catchment scale indication of dissolution and landscape erosion. A study of the river geochemistry of the Wujiang and Quingshuijiang rivers in the Guizhou Province of southern China (Han and Liu, 2004) demonstrated that dissolution in the catchments was attributable to both carbonic and sulphuric acids. Isotope studies ($^{87}Sr/^{86}Sr$) and the presence of nitrates indicate an anthropogenic source for a proportion of the sulphuric acid; thereby the implication is of anthropogenically accelerated erosion rates (43 to 49 mm per annum). Given the vulnerability of karst aquifers to contamination (Vesper et al., 2001), which results from the existence of open fast flow paths in the form of conduits, closely spaced monitoring in rivers can provide useful information with respect to point and diffuse sources and sinks of contaminants. A study of this type was undertaken by Banks and Palumbo-Roe (2010) in Rookhope Burn, a lead-zinc mining impacted karst catchment and tributary of the River Wear in northern England. This identified previously unrecognised resurgences of zinc-contaminated groundwater.

5. Geochemistry of secondary carbonate deposits

5.1 Secondary carbonate geochemistry

Secondary carbonate cements formed by meteoric water are representative of the local environmental conditions at the time of precipitation. Speleothems form in the vadose (unsaturated) zone by water dripping from the ceilings or walls of caves or from the overhanging edges of rock shelters. The precipitation of calcium carbonate is caused by the degassing of carbon dioxide into the cave, resulting in supersaturation of the groundwater with respect to calcium carbonate and consequential precipitation, which may also be biologically mediated. Seasonal lamination may occur as a consequence of seasonal variations in: drip rate (potential for variation in layer thickness); drip water supersaturation with respect to calcium carbonate, and cave climate (temperature, humidity and carbon dioxide concentrations). Lamination thickness measured in Belgian stalagmites by Genty and Quinif (1996) varied both between and within stalagmites, in the range 0.47 to 2.17 mm. Laminae were defined by variations in the density of the intercrystalline pores and inclusions. The pores were elongate parallel to the growth direction with their length ranging from 0.05 to 1 mm. Annual laminations formed as couplets of white porous laminae and dark compact laminae of palisade calcite. The dark laminae, which have a lower pore density, form during periods of moisture excess with insufficient time for surface degassing of carbon dioxide, resulting in run-off and precipitation on the sides of the speleothems. This causes an overall thinning of the laminae and localised widening of the stalagmite. During drier periods, the narrow white porous calcite laminae are attributable to a chemically efficient flow rate that allows sufficient time for carbon dioxide outgassing and precipitation of calcium carbonate at the top of the stalagmite.

Annual trace element variation within speleothems, in part attributable to colloid transport, provides valuable information that can be used to investigate climate change. Fairchild et al. (2001) used ion microprobe analyses of speleothems from five western European cave sites to demonstrate seasonality in concentrations of magnesium, strontium, barium, fluoride, hydrogen and phosphorus (as phosphate). The caves studied were: Crag Cave, County Kerry and Ballynamintra, County Wexford in Ireland; Uamh an Tartair, Sutherland, Scotland; Grotte Pere-Noël, Belgium, and Grotta di Ernesto, north-east Italy. It was established that: i) Magnesium, strontium and barium substitute directly for calcium. ii) Magnesium/ calcium ratios reflect the supply ratio, tending to be lower under high flow conditions, because of prior calcite precipitation along the flow path. iii) In cave environments with low sodium and magnesium/calcium ratios, and with a constant temperature, the key variables affecting strontium partitioning are the supply ratio of strontium/calcium and the growth rates, with more strontium being incorporated at higher flow rates. iv) The concentration of sodium and fluoride was found to reflect the growth rate (more sodium and fluoride at higher growth rates) rather than variations in the supply water chemistry, a consequence of the rapid incorporation of these elements, but results in charge imbalance (growth defect) that can be satisfied by the further incorporation of more trace elements and suggests greater concentration of trace elements during periods of rapid speleothem growth.

Comparable findings have been made in studies of tufa deposits which also comprise calcium carbonate that is representative of the local environmental conditions at the time of precipitation. This particularly applies to biologically moderated tufas, where the cyanobacteria do not have a direct influence on the rate, mineralogy, or geochemistry of the calcium carbonate that precipitates around them (Andrews and Brasier, 2005). The seasonality of tufas results from cyanobacterial blooms in the early spring, which facilitates nucleation of densely calcified darker layers. As the nutrient supply (diatoms) reduces, cyanobacterial growth slows and filaments aggregate into scattered bundles separated by cavities resulting in a lightly calcified, porous layer (Andrews and Brasier, 2005). Strontium, magnesium and manganese concentrations in summer precipitates have been shown to be higher than those in winter (Chafetz et al., 1991). Magnesium geochemistry is more complex than indicated by Chafetz et al. (1991); as with speleothems, upstream precipitation of calcite has been shown to have a minor effect on magnesium chemistry, and magnesium concentrations appear to be controlled by aquifer processes rather than temperature (Andrews and Brasier, 2005).

5.2 Stable isotopes (carbon and oxygen) – Environmental and hydrological implications

The ratio of $\delta^{18}O$ to $\delta^{16}O$ measured against a standard gives an indication of mean annual surface temperatures at the time of speleothem precipitation. Similarly, ratios of carbon isotopes may provide evidence for seasonality. Given no direct contact with the external atmosphere and constant temperature due to the limitation of air circulation, calcite precipitated in the resultant speleothem will form in isotopic equilibrium with the water from which it is precipitated. Changes in surface atmospheric conditions will be reflected in the percolating water and isotopically preserved in the speleothem and any fluid inclusions. However, not all speleothems can be studied in this way. For example, some may undergo diagenesis, or the speleothem may be contaminated by detrital matter (section 3.6).

Associated with the seasonality in tufa precipitation is a $\delta^{18}O$ trend to isotopically heavy winter-time tufa and isotopically light summer-time tufa. This is attributed to changing water temperature. $\delta^{13}C$ also shows seasonality with the summer-time sparry calcite exhibiting lower values than the micritic winter laminae. Matsuoka et al. (2001) suggested that seasonality in $\delta^{13}C$ in their Japanese study was attributable to winter degassing of carbon dioxide from groundwater fed spring water. In winter, the subsurface karst conduit air is warmer and less dense than the atmospheric air, so ventilation of the conduit air decreases the subsurface air P_{CO_2} and causes degassing of isotopically light $^{12}CO_2$ from the groundwater (Andrews and Brasier, 2005).

In a study of Quaternary pedogenic carbonates, designed to assess their usefulness as environmental indicators in Texas, Zhou and Chafetz (2010) undertook stable isotope analyses on a range of host strata. In the Gulf Coastal Plains, the Southern Highland Plains and the west, $\delta^{13}C$ values were found to vary in response to changes between C_4-dominated and C_3-dominated plants. Additionally, $\delta^{18}O$ values of the late Quaternary pedogenic carbonates decreased gradually from east to west, mimicking the spatial variation of $\delta^{18}O$ in modern meteoric water and increased distance from the Gulf of Mexico. However, the stable-isotope values developed on marine limestone and calcareous alluvium in central and southern Texas contained isotopic signatures that were derived from the host strata. These findings imply a variable potential for the interpretation of pedogenic carbonate stable isotope signatures.

5.3 Fractionation of isotopes

Fractionation of isotopes occurs where there is a change to the speleothem drip water chemistry prior to or during precipitation. This may occur in the soil, vadose zone, epikarst or in the cave environment. Potentially, fractionation may result in misinterpretation of isotope analyses. Fractionation effects can be estimated by analysing the variation in hydrogen isotopes, which are unaffected during crystal formation (Lauritzen, 1993). Additional information may be obtained from multi-proxy approaches, including non-conventional isotope systems, such as magnesium fractionation (Immenhauser et al., 2010). Fractionation can also occur as a consequence of post-depositional weathering of secondary carbonate deposits.

5.4 Isotopes and dating of secondary carbonate deposits

Radiometric dating methods are based on the decay of natural radio isotopes with a fixed decay constant (λ) for a given isotopic species. Cosmogenic isotopes are formed through cosmic reactions, e.g. with atmospheric nitrogen to generate carbon-14 (^{14}C), or rocks and soil (^{10}Be, ^{14}C, ^{26}Al and ^{36}Cl). The application of the appropriate techniques is based on the geochemistry and suspected age of the dateable material. The suspected age is considered because each of the isotopes has a different half-life e.g. 5730 +/- 40 years for ^{14}C. ^{14}C ages are generally determined from atom counts undertaken in an accelerator mass spectrometer. The principal dating technique used for secondary carbonate deposits is uranium-thorium dating. Radioactive uranium (^{238}U and ^{235}U) generates daughter isotopes ^{206}Pb and ^{207}Pb through radioactive decay (of α, β and γ particles) via intermediate daughter species including ^{234}U, and ^{230}Th. Uranium is soluble in carbonate waters; rock weathering results in the preferential removal of ^{234}U, whereas thorium is far less soluble and bonds preferentially to clay particles. Consequently, whilst ^{234}U may be incorporated in secondary carbonate cements, thorium will not be represented, thereby enabling dating from the U/Th ratio. For U/Th dating it is necessary that samples contain sufficient uranium with low ^{200}Pb concentrations. This dating technique has been extensively applied to speleothems (Baker et al., 2008; Richards and Dorale, 2003), tufas (albeit with the difficulty of low uranium concentrations in young tufas, Garnett et al., 2004) and increasingly to cemented rudites (Sharp et al., 2003). In interpreting the results, consideration needs to be given to the potential for sample contamination or for fractionation to have taken place, i.e. whether the system is open or closed to post depositional migration of constituent nuclides (Richards and Dorale, 2003). The use of other intermediate uranium daughter isotopes for dating, e.g. protactinium 231 (Edwards et al., 1987), forms the subject of ongoing research. Another research direction is the use of laser ablation linked ICPMS for determining uranium isotope ratios (Eggins et al., 2005). However, at present the precision of conventional TIMS or solution MC-ICPMS techniques has not been achieved. Other dating techniques include: optically stimulated luminescence (OSL), electron spin resonance (ESR; Grün, 1989) and amino acid racemisation. The latter is obtained from changes to the structure of organic matter, a process that is temperature controlled. Longer established dating techniques include palaeomagnetism (Rowe et al., 1988) and biostratigraphy, using fauna, flora and pollen.

Isotopic dating offers the potential for determining the minimum age of the surfaces (geological or anthropogenic) upon which the secondary carbonate deposits lie. For example, critical dating evidence (455 ka BP) has been derived from U/Th dating of tufa overlying an exposure of the Anglian Lowestoft Formation (a glacial till) at a Palaeolithic site at West Stow Suffolk, UK (Preece et al., 2007). At this locality the land snail assemblage identified in the tufa comprised woodland taxa indicative of a wetter and potentially warmer climate than the present day.

6. Secondary carbonate deposits as indicators of palaeohydrogeology, climate change and landscape evolution

6.1 Karstification and secondary carbonate deposition

The sources for secondary terrestrial carbonates are of interest to the karst hydrogeologist because of the potential to establish the palaeohydrological conditions at the time of precipitation (section 4.4). The spatial distribution of dissolution is of interest both to the hydrogeologist as an indicator of palaeo flow conditions and to the karst geomorphologist as it contributes to our understanding of landscape evolution.

6.2 Implications for interpreting climate change

The presence or absence of speleothems provides valuable information regarding sufficiency of water supply and soil carbon dioxide for their growth. As their growth is restricted to the vadose zone, they provide constraints on the opening of passages and water table fluctuations (Richards and Dorale, 2003). In the context of glacial-interglacial cycles, speleothem growth is at a maximum during warm periods. It is limited by frozen ground conditions, ice cover and subsequent melting. Frozen ground and lower biogenic carbon dioxide supply limit growth in permafrost conditions (Lowe and Walker, 1997). Growing from meteoric water, speleothems and tufa preserve a continuous record of climate change for the duration of their formation, albeit that this may be moderated by ground storage of the water. If certain conditions are satisfied, speleothems may contain annual laminae, generally of 10 to 1000 µm (Baker et al., 2008). The necessary conditions include: an annual cyclicity of the surface climate (e.g. seasonal monsoon, annual migration of the inter-tropical convergence zone, or a seasonal moisture deficit); a cyclic signal expressed in the speleothem as a consequence of groundwater, or cave atmosphere signal transfer, and a depth of speleothem such that groundwater moderation does not mask the climatic variation (Baker et al., 2008). Thus, detailed studies of annual growth laminae offer the potential to provide evidence of past precipitation, temperature, or atmospheric circulation. This can be done using a number of techniques (Table 3). Annual laminae, which are a function of palaeoclimate, define speleothem growth rate. Warmer, moister conditions are associated with increased growth, except under peat soils where warmer, drier conditions promote growth (Fairchild et al., 2001). Laminations represent a change in chemical composition of the speleothem that can be made visible by imaging or chemical mapping (Baker et al., 2008; Table 3). Visibility of the laminae depends on alternation of the crystal arrangement with a well-defined morphology. In some examples, e.g. Genty and Quinif (1996) there is an alternation between dark compact calcite and white porous calcite.

Providing laminae periodicity can be established and it can be demonstrated that the change in annual lamina thickness reflects surface climate variations through speleothem growth rate and geochemistry, then quantitative reconstruction of climate from stalagmite growth layers, or lamina-climatology (Baker et al., 2008) studies can be undertaken. Using this technique, Genty and Quinif (1996) distinguished 11-year cyclicity in speleothem growth in a pre-Holocene Belgian stalagmite, which they suspected of being linked to sunspot activity. Not all laminae are annual. Sub-annual laminae are more likely to be present in speleothems that grow closer to the ground surface without moderation by groundwater. Investigation of the periodicity of the laminae can be undertaken using: dating techniques; laminae counting between events; drip water chemistry monitoring (including fluorescence) or a comparison of observed with theoretical prediction of lamina width. Attention has also been given to the application of hyperspectral imaging, the opportunity for which lies in the potential to use the near infra-red range to map out areas of stalagmite where fluid inclusions are present. The sensitivity of speleothem growth to local conditions is such that there can be variation in drip chemistry within a single cave system, e.g. Pitty (1966) in Poole's Cavern, Derbyshire, UK. This results from the variation in individual flow paths and the consequential climate filtering, which may give rise to poor correlation between individual speleothems within the same cave system. This difficulty is best overcome using a multi-parameter investigative approach (Baker et al., 2008), e.g. lamina thickness, stable isotope investigation and trace element geochemistry.

Technique	Application	References
Measurement of change in laminae width using conventional transmission and reflected light microscopy on polished and thin sections. Scanning electron microscopy to look at pore spaces.	Speleothems from across the world including, Belgium; Brazil and Ethiopia.	Genty and Quinif, 1996; Soubiès et al., 2005, and Baker et al., 2007.
Studies of trapped pollen.	Speleothems.	McGarry and Caseldine, 2004.
Ratios of stable isotopes.	Speleothems, tufa, cements.	Andrews and Brasier, 2005; McDermott, 2004.
Calcite-aragonite couplets.	Speleothems, examples studied from north-west Botswana and Nepal.	Railsback et al., 1994, and Denniston et al., 2000.
Variations in trace element ratios.	Speleothems.	Fairchild et al., 2001.
Optical luminescence.	Speleothems; examples include Poole's Cavern, Derbyshire, U.K.	Baker et al., 1999.
Hyperspectral imaging used to discriminate laminar density contrasts.	Speleothems; examples studied from north-east Turkey.	Jex et al., 2008.

Table 3. Secondary carbonate imaging techniques.

Remnant organic acids trapped within the calcite growth rings can be investigated using luminescence studies and UV microscope technology. Fluorescence can be observed using reflected light microscopy to detect emission wavelengths of between 400 and 480 nm with a mercury light source to provide UV excitation at wavelengths of between 300 and 420 nm. Trace element and UV-fluorescent laminae studies provide information at a finer resolution (annual to subannual level) than the other techniques listed in Table 3, thereby offering the potential to research the changing nature of seasonality through time. During periods of higher or more intense precipitation, organic material can be washed on to the surface of speleothems. By applying these techniques to investigate growth laminae, conclusions can be drawn regarding the frequency of cyclonic storms and hence the changing atmospheric circulation patterns (Baker et al., 1999; 2008; Soubiès et al., 2005).

Stable isotope studies of speleothems and tufas are now well established proxies for interpreting climate change (Andrews, 2006; Baker et al., 2008; Andrews and Brasier, 2005; McDermott, 2004, Richards et al., 1994). Garnett et al. (2006) carried out stable isotope analyses of Late-glacial and early Holocene tufa deposits from Caerwys, North Wales, UK, demonstrating that Late-glacial tufa (pre-9000 years BP) probably formed during a period of climatic warming with summer water temperatures in the range 13 to 16.5°C, followed by a period of cooling associated with the cessation of tufa precipitation. Analysis of oxygen isotope ratios from pedogenic carbonates can also provide useful palaeoclimatic information, reflecting changes in the humidity and frequency of soil wetting (Lowe and Walker, 1997; Pendall et al., 1994).

Whilst laminar growth patterns have not been reported in secondary carbonate cements, more than one type or phase of cement can be present and these can be used to interpret depositional environments. For example, in a detailed petrological study carried out in conjunction with stable isotope analyses, Strong et al. (1992) identified a range of cement fabrics in carbonate cemented Late Devenesian glacial gravels.

6.3 Incision and the interpretation of groundwater base-levels

Speleothems have been used to determine sea level change, e.g. in the Bahamas (Gascoyne et al., 1979 and Richards et al., 1994). Dating of secondary carbonate deposits also offers the potential for determining rates of landscape incision. Where cemented rudites have formed on a geomorphic or anthropogenic surface, dating of the cement provides a minimum age for the host surface. This has been found to be particularly relevant to the dating of glacial tills, e.g. Preece et al. (2007). In the situation that a dated surface has been incised, the opportunity arises to calculate the rate of incision. It should be noted that the rate of incision may equate to either surface lowering, or to glaciogenic or tectonic uplift. Calculations of this type have been undertaken using dated tufa deposits in Lathkill Dale, Derbyshire, UK (Banks et al., 2011).

Similarly, although it is feasible that more than one cave level can be hydraulically operational at a given time, dating of speleothems has demonstrated that the oldest material is commonly found in the higher caves and that the onset of speleothem formation is progressively later in lower caves (Ford and Williams, 1989; Lowe and Walker, 1997). This may be a response to the progressive lowering of the water table and therefore another potential means of calculating rates of incision.

6.4 Carbon dioxide budgeting

The cycle of dissolution and precipitation of calcium carbonate involves carbon dioxide cycling. Consequently, bacterial carbonate precipitation in terrestrial environments may be crucial for atmospheric carbon dioxide budgeting (González-Muñoz et al., 2010).

6.5 Discriminating between point and diffuse contaminant sources

Monitoring the flux of dissolved species in streams in karst environments can provide useful information regarding the functioning of karst aquifers, in particular to determine sources and sinks of calcium carbonate. Associated with this, there is the potential to discriminate between diffuse and point source contaminants in the stream environment. Flow networks in karst aquifers facilitate the focusing of contaminants (Vesper et al., 2001). The flux associated with the latter has a more discrete, spiky impact on water quality, which is readily discernible with customised monitoring networks (Banks and Palumbo-Roe, 2010).

6.6 Dating of catastrophic events

Catastrophic events, such as earthquakes, can damage speleothems for which U-series dates can be obtained (Forti, 1997; Ford and Hill, 1999). Equally, the occurrence of geohazard events may be associated with secondary carbonate precipitation, e.g the Fern Pass rockslide, Austria (Ostermann et al., 2007). This provided new sources of carbonate and new flow paths, facilitating precipitation of secondary terrestrial carbonate deposits and providing a means of dating such events in the geological record.

7. Conclusions

This overview of secondary terrestrial carbonate deposition in karst environments demonstrates that the carbonates have a widespread distribution, display a variety of forms and result from a range of depositional and diagenetic processes. Such variability has resulted in numerous descriptive terms for the deposits as well as some complexities in classification. Examples cited in this chapter illustrate how detailed studies of the depositional setting, form and geochemistry of the carbonates can provide important information about formative processes, age and past environmental conditions (climatic, hydrogeological, hydrochemical, biological, geomorphological and anthropogenic). Such studies have particular relevance to interpretations of climate change and landscape evolution. Recent research has focused on the study of: terrestrial carbonate deposits in deep time (Brasier, 2011); diagenetic processes (Brasier et al., 2011; Martín–García et al., 2009); refining dating techniques for tufas and carbonate cements (Sharp et al., 2003; Garnett et al., 2004); dating using laser ablation linked ICPMS (Eggins et al., 2005); hyperspectral imaging of secondary carbonate deposits (Jex et al., 2008); furthering understanding of depositional processes, particularly the biological processes (Dittrick and Sibler, 2010; González–Muñoz et al., 2010, and Jones, 2010); cave atmosphere and drip chemistry (Baldini et al., 2006); collecting and interpreting continuous monitoring data (spring chemistry), and analysing the reasons for tufa decline, e.g. eutrophication (Forbes et al., 2010). The ongoing description and interpretation of "new" exposures of secondary carbonate deposits, e.g. Forbes et al. (2010) has the potential to underpin the conceptual understanding of the hydrogeology of these deposits.

8. Acknowledgements

This is published with the permission of the Executive Director of the British Geological Survey (NERC). The text has been significantly improved as a consequence of reviews by Dr Andrew Farrant and Dr Anthony Cooper at the British Geological Survey. Dr Cooper is thanked for the provision of Fig. 10.

9. Glossary

Activity constant:	The ratio of the apparent to the actual concentrations of ions that results from the inter-ionic forces of attraction in more concentrated solutions of a solute.
BP:	Before present.
Columnar:	Upright, elongate growth form.
Deamination:	Removal of an amine group (organic compound or functional group containing an N atom) from a molecule.
Dissociation constant:	Dissociation is a process whereby a single compound splits into two or more smaller products that can easily recombine to form the reactant. The dissociation constant reflects the extent of incomplete dissociation.

Doline:	Enclosed depression centred on a sink hole, or intercepted cave passage.
Epikarst:	Karst zone that lies closes to the surface, encompassing the soil and weathered bedrock.
EPS:	Extracellular polymeric substances, which are characterised by the presence of different proteins, uronic acids, pyruvic acid and sulphate groups.
ICPMS:	Inductively coupled plasma mass spectrometry.
Inception horizon:	Part of a rock succession that is particularly susceptible to the earliest cave forming processes.
Luminescence:	Light that usually occurs at low temperatures.
MC ICPMS:	Multi collector inductively coupled plasma mass spectrometry.
Mucopolysaccharide:	Polysaccharide containing an amino group.
Oncolite:	Carbonate encrusted nodules developed around a stone, or another nucleus (Tucker, 2011; Viles and Goudie, 1990).
Palisade calcite:	Calcite that is formed with elongate crystals perpendicular to the growth surface.
Picocyanobacteria:	Microscopic (0.2-2.0 µm) bacteria.
Polysaccharide:	Carbonate comprising a chain of sugars (monosaccharides) that be of more than one type.
Pyruvic acid:	An organic acid and ketone ($CH_2COCOOH$).
Sheath:	A bacterial sheath surrounds certain filamentous bacteria, particularly those in water.
Uronic acids:	A sugar with carbonyl and carboxylic acid function.
Speleogenesis:	Cave forming processes.
Stromatolites:	A term that is used to describe cemented algal mats from the coastal zone and is commonly applied to freshwater deposits of a similar form (Tucker, 2011; Viles and Goudie, 1990).
TIMS:	Thermal ionization mass spectrometry.
Twilight zone:	Portion of the cave that forms the transition between the unlit cave interior and the outside world (Jones, 2010).
Vaterite:	Metastable polymorph of calcium carbonate.

10. References

Aharon, P., Rasbury, M. and Murgulet, V. 2006. Caves of Niue Island, South Pacific: Speleothems and water geochemistry. In: Harmon, R.S. and Wicks, C. Editors. Perspectives on karst geomorphology, hydrology and geochemistry – A tribute volume to Derek C. Ford and William B. White: *Geological Society of America Special Paper*, 404, 283-295.

Andrews, J.E. 2006. Palaeoclimatic records from stable isotopes in riverine tufas: Synthesis and review. *Earth Science Reviews*, 75, 85-104.

Andrews, J.E., Pedley, H.M. and Dennis, P.F. 1994. Stable isotope record of palaeoclimatic change in a British Holocene tufa. *The Holocene*, 4, 349-355.

Andrews, J.E. and Brasier, A.T. 2005. Seasonal records of climatic change in annually laminated tufas: short review and future prospects. *Journal of Quaternary Science*, 20, 5, 411-421.

Appelo, C.A.J. and Postma, D. 2005. *Geochemistry, Groundwater and Pollution*. Balkema, Rotterdam. 536pp.

Atkinson, T.C. 1977a. Diffuse flow and conduit flow in limestone terrain in the Mendip Hills, Somerset (Great Britain). *Journal of Hydrology*, 35, 93-110.

Atkinson, T.C. 1977. Carbon dioxide in the atmosphere of the unsaturated zone: an important control of groundwater hardness in limestones. *Journal of Hydrology*, 35, 111-123.

Baker, A., Proctor, C.J., and Barnes, W.L. 1999. Variations in stalagmite luminescence laminae structure at Poole's Cavern, England, AD 1910-1996: calibration of a palaeoprecipitation proxy. *The Holocene*, 9, No. 6, 683-688.

Baker, A., Smith, C.L., Jex, C., Fairchild, I.J., Genty, D. and Fuller, L. 2008. Annually laminated speleothems: a Review. *International Journal of Speleology*, 37, 3, 193-206.

Baldini, J.U.L., McDermott, F. and Fairchild, I.J. 2006. Spatial variability in cave drip water hydrochemistry: Implications for stalagmite paleoclimate records. *Chemical Geology*, 235, 390-404.

Banks, V.J., Gunn, J. and Lowe, D.J. 2009. Stratigraphical influences on the limestone hydrogeology of the Wye catchment, Derbyshire. *Quarterly Journal of Engineering Geology and Hydrogeology*, 42, 211-225.

Banks, V.J. and Palumbo-Roe, B. 2010. Synoptic monitoring as an approach to discriminating between point and diffuse source contributions to zinc loads in mining impacted catchments. *Journal of Environmental Monitoring*, 12, 1684-1698.

Banks, V.J. Jones, P.F., Lowe, D.J., Lee, J.R., Rushton, J.C. and Ellis, M.A. 2011. Review of tufa deposition and palaeohydrological conditions in the White Peak, Derbyshire, UK: Implications for Quaternary landscape evolution. *Proceedings of the Geologists' Association* (in Press) 10.1016/j.pgeola.2011.03.011.

Bertenshaw, M. P. 1981. Hydrogeochemical indication of mineral deposits in the limestones of Derbyshire. *Transactions of the Institution of Mining and Metallurgy (Section : Applied Earth Science)*, 90, B167-B173.

Bischoff, J.L., Julia, R., Shanks, W.C. and Rosenbauer, R.J. 1994. Karstification without carbonic acid: Bedrock dissolution by gypsum-driven dedolomitization. *Geology*, 22, 995-998.

Brasier, A.T. 2011. Searching for travertines, calcretes and speleothems in deep time: Processes, appearances, predictions and the impact of plants. *Earth-Science Reviews*, 104, 213-239.

Brasier, A.T., Andrews, J.E. and Kendall, A.C. 2011. Diagenesis or dire genesis? The origin of columnar spar in tufa stromatolites of central Greece and the role of chironomid larvae. *Sedimentology, doi: 10.1111/j.1365-3091.2010.01208.x.*

Chafetz, H.S. and Folk, R.L. 1984. Travertines: depositional morphology and the bacterially constructed constituents. *Journal of Sedimentary Petrology*, 54, 289-316.

Chafetz, H.S., Utech, N.M. and Fitzmaurice, S.P. 1991. Differences in the $\delta^{18}O$ and $\delta^{13}C$ signatures of seasonal laminae comprising travertine stromatolites. *Journal of Sedimentary Geology*, 162, 199-218.

Chen, J., Zhang, D.D., Wang, S., Xiao, T. and Huang, R. 2004. Factors controlling tufa deposition in natural waters at waterfall sites. *Sedimentary Geology*, 166, 353-366.

Clarke, I and Fontes, J.C. 1990. Palaeoclimatic reconstruction in Northern Oman based on carbonates from hyperalkaline ground-waters. *Quaternary Research*, 33, 320-336.

Curtis, M.H. 2002. Pedogenic carbonate: links between biotic and abiotic $CaCO_3$. 17TH WCSS, 14-21 August, 2002, Thailand. Paper 897, 1-9.

Denniston, R.F., Gonzalez, L.A., Asmerom, Y., Sharma, R.H. and Reagan, M.K. 2000. Speleothem evidence for changes in Indian summer monsoon precipitation over the last 2300 years. *Quaternary Research,* 53, 196-202.

Dipova, N. 2011. The engineering properties of tufa in the Antalya area of SW Turkey. *Quarterly Journal of Engineering Geology and Hydrogeology,* 44, 123-134.

Dittrich, M. and Sibler, S. 2010. Calcium carbonate precipitation by cyanobacterial polysaccharides. In: Pedley,.M. and Rogerson, M. (Eds). 2010. *Tufas and Speleothems: Unravelling the Microbial and Physical Controls.* Geological Society, London, Special Publication, 336, 51-63.

Downing, R.A. 1967. *The geochemistry of groundwaters in the Carboniferous Limestone in Derbyshire and the East Midlands.* Bulletin No. 27. *Geological Survey of Great Britain,* 289-307.

Drake, J.J. and Harmon, R.S. 1973. Hydrochemical environments of carbonate terrains. *Water Resources Research,* 9, No. 4, 949-957.

Edmunds, W.M. 1971. *Hydrogeochemistry of groundwaters in the Derbyshire Dome with special reference to trace constituents.* IGS Report 71/7.

Edwards, R.l., Chen, J.H., Wasserburg, G.J. 1987. ^{238}U-^{234}U-^{230}Th-^{232}Th systematics and precise measurement of timeover the past 500, 000 years. *Earth and Planetary Science Letters* 81, 175-192.

Eggins, S.M., Grün, R., McCulloch, M.T. Pike, A.W.G., Chappell, J., Kinsley, L., Mortimer, G., Shelley, M., Murray-Wallace, C.V., Spötl, C. and Taylor, L. 2005. In situ U-series dating by laser-ablation multi-collector ICPMS: new prospects for Quaternary geochronology. *Quaternary Science Reviews,* 24, 2523-2538.

Faccenna, C., Soligo, M., Billi, A., De Filippis, L., Funiciello, R., Rossetti, C. and Tuccimei, P. 2008. Late Pleistocene depositional cycles of the Lapis Tiburtinus travertine (Tivoli, Central Italy): Possible influence of climate and fault activity. *Global and Planetary Change,* 63, 299-308.

Fairchild, I.J., Baker, A., Borsato, A., Frisia, S., Hinton, R.W., McDermott, F. and Tooth, A. 2001. Annual to sub-annual resolution of multiple trace-element trends in speleothems. *Journal of the Geological Society, London* 158, 831-841.

Fetter, C.W. 2001. *Applied Hydrogeology.* 4 th Ed. Prentice-Hall, Inc. 598 pp.

Forbes, M., Vogwill, R. and Onton, K. 2010. A characterisation of the coastal tufa deposits of south-west Western Australia. *Sedimentary Geology,* 232, 52-65.

Ford, D.C. and Williams, P.W. 1989. *Karst Geomorphology and Hydrology.* Unwin Hyman, London.

Ford, D.C. and Hill, C.A. 1999. Dating of speleothems in Kartchner Caverns, Arizona. *Journal of Cave and Karst Studies,* 6, 2, 84-88.

Ford, D.C. and Williams, P.W. 2007. *Karst Hydrogeology and Geomorphology.* John Wiley and Sons Limited. 562pp.

Ford, T.D. and Pedley, H.M. 1996. A review of tufa and travertine deposits of the world. *Earth Science Reviews,* 41, 117-175.

Forti, P. 1997. Speleothems and Earthquakes. In Hill, C. and Forti, P.(Editors) *Caves and minerals of the world.* National Speleological Society, Huntsville, 284-285.

Francis, J.E. 1984. The seasonal environment of the Purbeck (upper Jurassic) fossil forests. *Palaeogeography, Paleoclimatology, Palaeoecology,* 48, 2-4, 285-307.

Fuller, L., Baker, A., Fairchild, I.J., Spötl, C., Marca-Bell, A., Rowe, P. and Dennis, P.F. 2008. Isotope hydrology of dripwaters in a Scottish cave and implications for stalagmite palaeoclimate research. *Hydrology and Earth System Sciences Discussions,* 5, 547-577.

Garnett, E.R., Gilmour, M.A., Rowe, P.J., Andrews, J.E. and Preece, R.C. 2004. ^{230}Th/^{234}U dating of Holocene tufas: possibilities and problems. *Quaternary Science Reviews,* 23, 947-959.

Garnett, E.R., Andrews, J.E., Preece, R.C. and Dennis, P.F. 2006. Late-glacial and early Holocene climate and environment from stable isotopes in Welsh tufa. *Quaternaire,* 17 (2), 31-42.

Gascoyne, M., Benjamin, G.J. Schwarcz, H.P. and Ford, D.C. 1979. Sea level lowering during the Illinoian glaciation: evidence from a Bahama "blue hole". *Science,* 205, 806-808.

Genty, D. and Quinif, Y. 1996. Annually laminated sequences in the internal structure of some Belgian stalagmites-importance for palaeoclimatology. *Journal of Sedimentary Research,* 66, No. 1, 275-288.

Glover, C. and Robertson, A.H.F. 2003. Origin of tufa (cool-water carbonate) and related terraces in the Antalya area, SW Turkey. *Geological Journal,* 38, 329-358.

González-Muñoz, M.T., Rodriguez-Navarro, C., Martĩnez-Ruiz, F., Arias, J.M, Merroun, M.L. and Rodriguez-Gallego, M. 2010. Bacterial biomineralization: new insights from Myxococcus-induced mineral precipitation. In: Pedley, .M. and Rogerson, M. (Eds). 2010. *Tufas and Speleothems: Unravelling the Microbial and Physical Controls.* Geological Society, London, Special Publication, 336, 31-50.

Goudie, A.S., Viles, H.A. and Pentecost, A. 1993. The late-Holocene tufa decline in Europe. *The Holocene,* 3, 181-186.

Grün, R. 1989. Electron spin resonance (ESR) dating. *Quaternary International,* 1, 65-109.

Gunn, J. 1981. Limestone solution rates and processes in the Witomo district, New Zealand. *Earth Surface Processes and Landforms,* 6, 427-45.

Gunn, J. 1986. Solute processes and karst landforms. In *Solute Processes.* Ed. Trudgill, S.T. John Wiley and Sons Ltd. pp. 363-437.

Han, G. and Liu, C-G. 2004. Water geochemistry controlled by carbonate dissolution: a study of the river waters draining karst-dominated terrain, Guizhou Province, China. *Chemical Geology,* 204, 1-21.

Hill, C.A. and Forti, P. 1995. The classification of cave minerals and speleothems. *International Journal of Speleogenesis* 24, 1-4, 77-82.

Immenhauser, A., Buhl, D., Richter, D., Niedermayr, A., Riechelmann, D., Dietzel, M. and Schulte, U. 2010. Magnesium-isotope fractionation during low-Mg calcite precipitation in a limestone cave-Field study and experiments. *Geochimica et Cosmochimicia Acta,* 74, 4346-4364.

Jacobson, R.L. and Langmuir, D. 1974. Controls on the quality variations of some carbonate spring waters. *Journal of Hydrology,* 23, 247-265.

Jex, C., Claridge, E, Baker, A. and Smith, C. 2008. Hyperspectral imaging of speleothems. *Quaternary International,* 187, 5-14.

Jones, B. 2010. Microbes in caves: agents of calcite corrosion and precipitation. In: Pedley,.M. and Rogerson, M. (Eds). 2010. *Tufas and Speleothems: Unravelling the Microbial and Physical Controls.* Geological Society, London, Special Publication, 336, 7-30.

Kaufmann, G. and Dreybrodt, W. 2007. Calcite dissolution kinetics in the system $CaCO_3$-H_2O-CO_2 at high undersaturation. *Geochimica et Cosmochimica Acta*, 71, 1398-1410.

Kawai, T, Kano, a., Matsuoka, J., and Ihara, T. 2006. Seasonal variation in water chemistry and depositional processes in a tufa-bearing stream in SW-Japan, based on 5 years of monthly observations. *Chemical Geology*, 232, 33-53.

Kendall, A.C. and Broughton, P.L. 1978. Origin of fabrics in speleothems composed of columnar calcite crystals. *Journal of Sedimentary Research*, 48, 2, 519-538.

Kehew, A.E. 2001. Applied chemical hydrogeology. Prentice Hall. 368 pp.

Lauritzen, S.E. 1993. Natural environmental change in karst: the Quaternary record. In Williams, P.W. (Editor) Karst Terrains: Environmental Changes and Human Impact. *Catena Supplement*, 25, Catena Verlag, Cremlingen-Destedt, Germany, 21-40.

Leybourne, M.I., Betcher, R.N., McRitchie, W.D., Kaszycki, C.A. and Boyle, D.R. 2009. Geochemistry and stable isotopic composition of tufa waters and precipitates from the Interlake Region, Manitoba, Canada: Constraints on groundwater origin, calcitization, and tufa formation. *Chemical Geology*, 260, 3-4, 221-233.

Liu, Z., Li, Q., Sun, H. and Wang, J. 2007. Seasonal, diurnal and storm-scale hydrochemical variations of typical epikarst springs in subtropical karst areas of SW China: Soil CO_2 and dilution effects. *Journal of Hydrology*, 337, 207-223.

Lorah, M.M. and Herman, J.S. 1988. Chemical evolution of a Travertine Depositing Stream. *Water Resources Research*, 24, No.9, 1541-155.

Love, K.M. and Chafetz, H.S. 1988. Diagenesis of laminated travertine crusts, Arbuckle Mountains, Oklahoma. *Journal of Sedimentary Research*, 58, 441-445.

Lowe, D.J. 2000. Role of stratigraphic elements in speleogenesis: the speleoinception concept. In: *Speleogenesis evolution of karst aquifers*. Eds. Klimchouk, A.B., Ford, D.C., Palmer, A.N. and Dreybrodt, W. pp. 65-76.

Lowe, J.J. and Walker, M.J.C. 1997. Reconstructing Quaternary Environments. Second Edition. Pearson Prentice Hall 446 pp.

Manning, D.A.C. 2008. Biological enhancement of carbonate precipitation: passive removal of atmospheric CO_2. *Mineralogical Magazine*, 72 (2), 639-649.

Martín-García, R., Alonso-Zarza, A.M. and Martín-Pérez, A. 2009. Loss of primary texture and geochemical signatures in speleothems due to diagenesis: Evidences from Castañar Cave, Spain. *Sedimentary Geology*, 221, 141-149.

Matsuoka, J., Kano, A., Oba, T., Watanabe, T., Sakaï, S. and Seto, K. 2001. Seasonal variation of stable isotopic compositions recorded in laminated tufa, SW Japan. *Earth and Planetary Science Letters*, 192, 31-44.

McDermott, F. 2004. Paleo-climate reconstruction from stable isotope variations in speleothems: a review. *Quaternary Science Reviews*, 23, 901-918.

McGarry, S.F. and Caseldine, C. 2004. Speleothem palynology: an undervalued tool in Quaternary studies. *Quaternary Science Reviews*, 23, 2389-2404.

Ostermann, M., Sanders, D., Prager, C. and Kramers, J. 2007. Aragonite and calcite cementation in "boulder controlled" meteoric environments on the Fern Pass rockslide (Austria): implications for radiometric age dating of catastrophic mass movements. *Facies*, 53, 189-208.

Palmer, A.N. 1991. Origin and morphology of caves. *Geological Society of America Bulletin*, 103, 1-21.

Palmer, A.N. 2002. Speleogenesis in carbonate rocks. In: *Evolution of Karst: From Prekarst to Cessation*. Gabrovsek, F.(Editor). Postojna. pp. 43-59.

Pazdur, A., Pazdur, M.F. and Szulc, J. 1988. Radiocarbon dating of Holocene calcareous tufa in Southern Poland. *Radiocarbon*, 30, 133-151.

Pedley, H.M. 1990. Classification and environmental models of cool freshwater tufas. *Sedimentary Geology*, 68, 143-154.

Pedley, M.H. and Rogerson, M. 2010. Introduction to tufas and speleothems. In: Pedley, .M. and Rogerson, M. (Eds). 2010. *Tufas and Speleothems: Unravelling the Microbial and Physical Controls*. Geological Society, London, Special Publication, 336, 1-5.

Pendall, E.G., Harden, J.W., Trumore, S.E. and Chadwick, O.A. 1994. Isotopic approach to soil carbonate dynamics and palaeoclimatic interpretations. *Quaternary Research*, 42, 60-71.

Pentecost, A. 1978. Blue-green algae and freshwater carbonate deposits. *Proceedings of the Royal Society of London B* 200, 43-61.

Pentecost, A. 1988. Growth and calcification of the Cyanobacterium *Homoeothrix crustacea*. *Journal of General Microbiology*, 134, 2665-2671.

Pentecost, A. 1991. Calcification processes in algae and cyanobacteria. In: Riding, R. (ed) Calcareous *Algae and Stromatolites*. Springer, Berlin, 3-20.

Pentecost, A. 1995. The Quaternary travertine deposits of Europe and Asia Minor. *Quaternary Science Reviews*, 14, 1005-1028.

Pentecost, A. 1999. The origin and development of the travertines and associated thermal waters at Matlock Bath, Derbyshire. *Proceedings of the Geologists' Association*, 110, 217-232.

Pentecost, A. and Lord, T. 1988. Postglacial tufas and travertines from the Craven district of Yorkshire. *Cave Science* 15, 15-19.

Pentecost, A. and Viles, H.A. 1994. A review and reassessment of travertine classification. *Géographie physique et Quaternaire*, 48, 3, 305-314.

Pitty, A.F. 1966. *An Approach to the Study of Karst Water*, 59 (University of Hull Occasional Papers in Geography, 5).

Pitty, A.F. 1968. The scale and significance of solutional loss from the limestone tract of southern Pennines. *Proceedings of the Geologists' Association*, 79 (2), 153-177.

Preece, R.C., Parfitt, S.A., Bridgland, D.R., Lewis, S.G., Rowe, P.J., Atkinson, T.C., Candy, I., Debenham, N.C., Penkman, K.E.H., Rhodes, E.J., Schwenninger, J-L., Griffiths, H.I., Whittaker, J.E. and Gleed-Owen, C. 2007. Terrestrial environments during MIS 11: evidence from the Palaeolithic site at West Stow, Suffolk, UK. *Quaternary Science Reviews*, 26, 1236-1300.

Railsback, L.B., Brook, G.A., Chen, J. 1994. Environmental controls on the petrology of a late Holocene speleothem from Botswana with annual layers of aragonite and calcite. *Journal of Sedimentary Research*, 64, 147-155.

Richards, D.A. and Dorale, J.A. 2003. Uranium-series Chronology and Environmental Applications of Speleothems. In: *Uranium-series Geochemistry* (Editors B Bourdon, G.M. Henderson, C.C. Lundstrom and S.P. Turner). *Reviews in Mineralogy and Geochemistry*, 52, 407-461.

Richards, D.A., Smart, P.L. and Lawrence Edwards, R. 1994. Maximum sea levels for the last glacial period from U-series ages of submerged speleothems. *Nature*, 367, 357-360.

Richardson, T.D. 1968. The use of chemical analysis of cave waters as a method of water tracing and indicator of type of strata traversed. *Transactions of the Cave Research Group of Great Britain,* 10, 2, 61-72.

Riding, R. 2000. Microbial carbonates: the geological record of calcified bacterial-algal mats and biofilms. *Sedimentology Supplement 1,* 47, 179-214.

Rieuwerts, J.H. 2000. *Lathkill Dale Derbyshire its Mines and Miners.* Landmark Publishing Limited. 110pp.

Rowe, P., Austin, T., and Atkinson, T. 1988. The Quaternary evolution of the British South Pennines from Uranium series and Palaeomagnetic data. *Annales de la Société géologique de Belgique,* T. 111, 97-106.

Selim, H.H. and Yanik, G. 2009. Development of the Cambazh (Turgutlu/MANISA) fissure-ridge-type travertine and relationship with active tectonics, Gediz Graben, Turkey. *Quaternary International,* 199, 157-163.

Sharp, W.D. Ludwig, K.R., Chadwick, O.A., Admundson, R. and Glaser, L.L. 2003. Dating fluvial terraces by ^{230}Th/U on pedogenic carbonate, Wind River Basin, Wyoming. *Quaternary Research,* 59, 139-150.

Sherwin, C.M. and Baldini, J.U.L. 2011. Cave air and hydrological controls on prior calcite precipitation and stalagmite growth rates: Implications for palaeoclimate reconstructions using speleothems. *Geochimica et Cosmochimica Acta,* 75, 3915-3929.

Shuster, E. T. and White, W. B. 1971. Seasonal fluctuations in the chemistry of limestone springs: a possible means for characterizing carbonate aquifers. *Journal of Hydrology,* 14, 93-128.

Şimşek, S. 1993. Karstic hot water aquifers in Turkey. In Hydrogeological Processes in Karst Terranes (Proceedings of the Antalya Symposium and Field Seminar, October, 1990; IAHS Publication 207).

Smith, D.I. and Atkinson, T.C. 1976. Process, landforms and climate in limestone regions, 367-409. In: Derbyshire, E. Ed. *Geomorphology and Climate,* 512pp.

Soubiès, F., Seidel, A., Mangin, A., Genty, D., Ronchail, J., Plagnes, V., Hirooka, S. and Santos, R. 2005. A fifty-year climatic signal in three Holocene stalagmite records from Mato Grosso, Brazil. *Quaternary International,* 135, 115-129.

Strong, G.E., Giles, J.R.A. and Wright, V.P. 1992. A Holocene calcrete from North Yorkshire, England: implications for interpreting palaeoclimates using calcretes. *Sedimentology,* 39, 333-347.

Sweeting, M.M. 1966. The weathering of limestones, with particular reference to the Carboniferous Limestones of northern England. In: Dury, G.H. (Editor) *Essays in Geomorphology.* Heinemann. London. pp. 177-210.

Taylor, D.M., Griffiths, H.I., Pedley, H.M. and Prince, I. 1994. Radiocarbon-dated Holocene pollen and ostracod sequences from barrage tufa-dammed fluvial systems in the White Peak, Derbyshire, U.K. *The Holocene,* 4, No.4, 356-364.

Terjesen, S.G., Egra, O., Thoresen, G. and Ve, A. 1961. Phase boundary processes as rate determining steps in reactions between solids and liquids. The inhibitory action of metal ions on the formation of calcium bicarbonate by the reaction of calcite with aqueous carbon dioxide. *Chemical Engineering Science,* 14, 277-289.

Thrailkill, J. 1968. Chemical and Hydrologic Factors in the Excavation of Limestone Caves. *Geological Society of America Bulletin,* 79, 19-46.

Tooth, A.F. and Fairchild, I.J. 2003. Soil and karst aquifer hydrological controls on the geochemical evolution of speleothem-forming drip waters, Crag Cave, southwest Ireland. *Journal of Hydrology*, 273, 51-68.

Tucker, M.E. 2011. *Sedimentary rocks in the Field: a practical guide*. Wiley Blackwell. 275 pp.

Tucker, M.E. and Wright, P.V. 1990. Carbonate Sedimentology. Blackwell Science Ltd. 482pp.

Vervier, P. 1990. Hydrochemical characterization of the water dynamics of a karstic system. *Journal of Hydrology*, 121, 103-117.

Vesper, D.J., Loop, C.M. and White, W.B. 2001. *Contaminant transport in karst aquifers. Theoretical and Applied Karstology*, 13-14, 101-111.

Viles, H.A. and Goudie, A.S. 1990. Tufas, travertines and allied carbonate deposits. *Progress in Physical Geography*, 14, 19-41.

Viles, H.A. and Pentecost, A. Tufa and Travertine. In: Nash, D.J. and McLaren, S.J. (Editors) 2007. Geochemical sediments and landscapes. Blackwell Publishing. pp. 173-199.

White, W.B. 2002. Karst hydrology: recent developments and open questions. *Engineering Geology*, 65, 85-105.

Woo, K.S. and Choi, D.W. 2006. Calcitization of aragonite speleothems in limestone caves in Korea: Diagenetic process in a semiclosed system. In: Harmon, R.S. and Wicks, C. Editors. Perspectives on karst geomorphology, hydrology and geochemistry – A tribute volume to Derek C. Ford and William B. White: *Geological Society of America Special Paper*, 404, 297-306.

Worthington, S.R.H. 1991. *Karst Hydrogeology of the Canadian Rocky Mountains*. Unpublished Ph.D. thesis. McMaster University. 370pp.

Worthington, S.R.H. and Ford, D.C. 1995. High Sulfate concentrations in limestone springs: An important factor in conduit initiation? *Environmental Geology*, 25, 9-15.

Zhou, J. and Chafetz, H.S. 2010. Pedogenic Carbonates in Texas: Stable-Isotope Distributions and Their Implications for Reconstructing Region-Wide Paleoenvironments. *Journal of Sedimentary Research*, 80, 2, 137-150.

3

Significance of Hydrogeochemical Analysis in the Management of Groundwater Resources: A Case Study in Northeastern Iran

Gholam A. Kazemi[1] and Azam Mohammadi[2]

[1]*Faculty of Earth Sciences, Shahrood University of Technology, Shahrood*
[2]*Senior Hydrogeologist, Base Studies Section of the Water Resources Department,*
Northern Khorasan Regional Water Company, Bojnourd
Iran

1. Introduction

Hydrogeochemistry is a sub-discipline of hydrogeology which is referred to as Chemical hydrogeology in some references (e.g. Domenico and Schwartz, 1990) and Groundwater geochemistry in some others (e.g. Merkel and Planer-Friedrich, 2005). One may also points to "Contaminant hydrogeology" used by Fetter (1998) as another term that carries the same syllabus as Hydrogeochemistry. In all fields of science and engineering, sub-disciplines are developed to accomplish some missions, fulfill some requirements and supplement the base subject. The same is true with hydrogeochemistry. This sub-discipline has been developed to deal with quality, contamination, chemistry, chemical processes and reactions that take place in various groundwater systems. Due to the importance of water quality issues, this sub-discipline has gradually changed into a well established field of research. For instance, Prasanna et al. (2011) clearly demonstrates that the study of quantity of water alone is not sufficient to solve the water management problems because its uses for various purposes depend on its quality. In addition, a number of other researchers show that the hydrogeochemical characteristics of groundwater and groundwater quality in different aquifers over space and time are important parameters in solving the groundwater management issues (Panigrahy et al., 1996; Atwia et al., 1997; Ballukraya and Ravi, 1999; Ramappa and Suresh, 2000). At the start, it should be pointed out that the quality of groundwater depends on the chemical composition of recharge water, the interaction between water and soil, soil–gas interaction, the types of rock with which it comes into contact in the unsaturated zone, the residence time of groundwater in the subsurface environment and the reactions that take place within the aquifer (Freeze and Cherry, 1979; Hem, 1989; Appelo and Postma, 2005). Hydrogeochemical processes such as dissolution, precipitation, absorption and desorption, ion exchange reactions and the residence time along the flow path which controls the chemical composition of groundwater, constitutes the other issues that are dealt with in hydrogeochemistry.

2. Framework of a hydrogeochemical study

There is a routine working flow chart for a hydrogeochemical study. It starts with preliminary desk studies including data collection and collation. Once all the available data-related to the aquifer in question- are assembled and assessed, data gaps are identified and deficiencies are brought into attention. The type of the data that needs to be created and the costs involved are all carefully evaluated. Often, the missing and the necessary data we think about at the first glance are the chemistry data. To gather these, groundwater sampling for chemical analysis are proposed and meticulously planned. Depending on the financial constraints and the level of precision required to solve the problem in hand, the number of samples varies. Sometime, it might be needed to drill new piezometers for this purpose. Before any sampling campaign, necessary equipments such as bottles, additives including acids and reagents, field probes, filter papers, pumps and bailers, various size hoses, isolating box or car fridge and a variety of other stuff- depending on the specific aim of the study- should be prepared. Field vehicle, maps, food and proper shoe and clothing are also needed as is the case with any geological field investigations. Collected groundwater samples are then transported to the relevant laboratory in due time, where they are analyzed using methods specified for different types of chemical elements, anthropogenic constituents, microbes and bacteria. The quality control techniques (duplicated samples, comparing the results which were produced by different laboratories analyzing the same samples, etc) should be undertaken to ensure reliable data. Working with various computer codes to interpret the data and graphically illustrate the results, is the program of work for the next step. At the final stage of the study, hydrogeochemical findings should be compared with the other geological, hydrogeological and field observations results and either revised accordingly or further filed investigation and water sampling be planned.

3. Applications of hydrogeochemical analysis

What we learn from a hydrogeochemical study? What are the objectives of a hydrogeochemical analysis program? Hydrogeochemical investigations are generally carried out to:

1. Identify the cause of anthropogenic contamination and the source of natural salinity
2. Evaluate the fate of the pollutants in the subsurface environment
3. Calculate the time needed to flush out the contaminants and clean the aquifer
4. Pinpoint the type of the geochemical reactions that are taking place within an aquifer system
5. Asses the quality of groundwater for different uses
6. Identify the factors governing the chemical composition of groundwater and the contribution made by each factor
7. Map the spread and the three dimensional picture of the contamination plume
8. Constrain the extent of the saltwater intrusion zone
9. Quantify leakage between surface water and groundwater and locate the hyperhoic zone
10. Compute the natural capacity of the aquifer in attenuating the pollutants

11. Study salinity stratification in groundwater systems and likeliness of variable density flows.

4. A hydrogeochemical case study in Northeastern Iran

Iran is located in a semi-arid area with an annual precipitation rate of 250 mm/year which is less than one third of that of the world average. In the arid and semi-arid environments with limited water resources such as Iran, groundwater constitutes a significant part of the available water resources (Subyani, 2005). Groundwater provides more than half of the total annual water demand in Iran; the uncontrolled groundwater use accompanied by successive draughts in the recent years has adversely affected the quality and quantity of Iran's subsurface waters. This is particularly true in central plateau where high temperature and low precipitation rates prevail. Furthermore, in such environments, groundwater chemistry evolves rapidly and the salinity increases swiftly, leading to restrictions in water utilization and limitations on the development and management of unconfined alluvial aquifers (Subyani, 2005). Several factors lead to rise in the salinity of groundwater. Some of these are local, such as hydrogeological conditions, rate of natural recharge and irrigation, while others are regional in nature. The latter group includes the aridity of the environment and the irregular and unpredictable occurrence of atmospheric precipitation (Eagelson, 1978 and 1979).

5. Description of the study area

Safi Abad watershed in northeastern Iran is located in the south of the Northern Khorasan province, one of the 31 provinces which make up the land of the country (Figure 1). It occupies an area of 2,593 Km², 848 Km² of which is in the form of a plain (Safi Abad plain) and the rest is highlands and mountains. There is no large city in the area and the largest village, Safi Abad, lies in the southern part of the watershed. All together, there are 77 villages in this watershed, the population of which reach 21,152 people. The climate of the Safi Abad watershed is classified as arid and semi-arid with an average annual precipitation of 241 mm. The local temperature rises above 40 °C in hot summer days and it sharply drops to a few degrees below freezing point during the clod winter days. The only local permanent river is Kale Shoor River which flows along east-west direction and is adjourned by two main tributaries. It originates from the mountains in the northern and south eastern parts of the Safi Abad plain and exit from the north western part after draining the area. Due to the scarcity of the surface water resources, water demand for agricultural and potable purposes is supplied by a number of various sizes qanats, springs and water wells mainly mainly in the northern part of the watershed (Note that qanat is a traditional structure, a horizontal gallery which yields groundwater by gravity). Agriculture and animal husbandry are the two main job markets in the region; cotton, a variety of crops, sugar bits, zirah (an Iranian costly product used as flavor in tea and food) and nuts are the main agricultural products. Natural vegetation covers in the highlands consists mainly of short trees and sparsely spaced tall trees.

In terms of regional geology, Safi Abad watershed is a part of the Zone of Central Iran, the oldest rock outcrops of which are Paleozoic in age (Stocklin, 1968). Local geological map of the study area is shown in Figure 2. As it can be seen from this figure, various lithologies

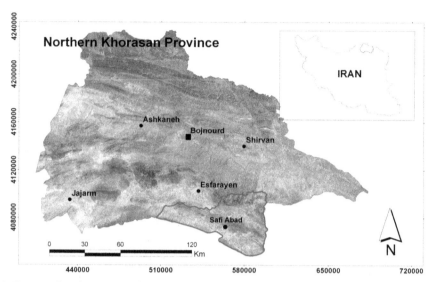

Fig. 1. Geographical position of the study area in the southern part of the Northern Khorasan Province, northeastern Iran.

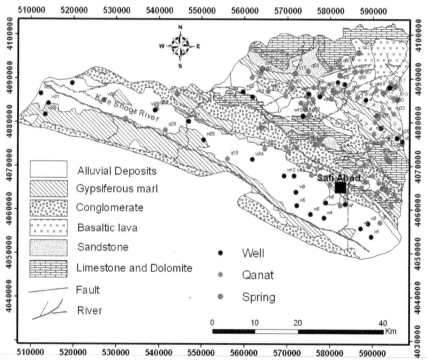

Fig. 2. Geological map of the Safi Abad Watershed (Based on Afshar Harb et al., 1987) and the location of the groundwater samples. Samples from wells, qanats and springs are differentiated.

such as limestone, dolomite, sandstone, shale-sandstone-gypsum sequence and basaltic lava underlie the area. The diverse range of rocks and sediments has led to different erosional pattern in the area. Unconsolidated and easily erodable parts of the plain are covered by marl and conglomerate which make up both the lowlands and the extensive valley floors. In contrast, highlands are formed by massive limestones and Eocene volcanic rocks which are over 2,000 m in height in some points. The minimum elevation in the study area is 1,050 m at the outlet of Kale Shoor river and the maximum elevation is 3,034 m above sea level.

6. Groundwater sampling

In the study area, groundwater utilization takes place through water wells, springs and qanats. There are 133 water wells, 190 qanats and 125 springs in the area which produce 51, 16.4 and 21 MCM (Million Cubic Meters) of groundwater per year, respectively; groundwater production rate is therefore 88.4 MCM per annum in an average year. Wells are mostly located in the plains while the majority of qanats and springs emerge from the highlands and mountains (Figure 2). Like everywhere, more than 95 percents of the extracted groundwater is used in agriculture sector; the proportional contribution of wells, qanats and springs is slightly different, however. A total of 35 water wells, 22 springs and 60 qanats were sampled from different parts of the aquifer underlying Safi Abad watershed in 2010. The sampling sites, altogether 117 sites, are shown in Figure 2. Electrical conductivity (EC) and hydrogen ion concentration (pH) were measured in the field using portable EC and pH meters. Water samples were then analyzed in the laboratory for acidity, electrical conductivity and major ions using standard methods. The summary of the results of the chemical analysis of groundwaters and the calculated charge balance error are presented in Table1. Low values of charge balance errors demonstrate that the accuracy of the analysis is within the acceptable range.

	EC µS/cm	pH	Anions meq/L				Cations meq/L				% error
			Na^+	K^+	Mg^{2+}	Ca^{2+}	HCO_3^-	CO_3^{2-}	SO_4^-	Cl^-	
Average	3322	7.84	21.4	0.599	6.63	5.83	4.68	0.072	10.83	19.2	-0.75
Minimum	274	6.2	0.2	0	0.2	0.8	1.5	0	0.3	0.3	-6.25
Maximum	35825	8.8	261	14.8	47	77	9.5	0.9	93	285	3.9

Note that percent error values are statistically calculated, otherwise -6.27 is the highest error.

Table 1. Summary of physico-chemical data of the water samples (Number of samples: 117).

7. Results and discussions

7.1 Electrical conductivity and pH

The EC of the samples ranges widely from 274 µS/cm to 35,825 µS/cm, with an average value of 3,322 µS/cm. Forty nine samples have ECs less than 1000, 23 samples less than 2000, 19 samples over 2000 and less than 5000 and the remaining 26 samples have ECs higher than 5000 μS/cm. This is a clear indication that the aquifer in question has been subjected to salinization processes either naturally or anthropogenicly. Saline samples are mostly from the plain and from the wells. Samples pH values are all but one above neutrality which reflects the basic nature of the groundwater. Figure 3 shows that there is a slight negative correlation between EC and pH values, most probably due to high sodium content of these samples as it is demonstrated in the next section.

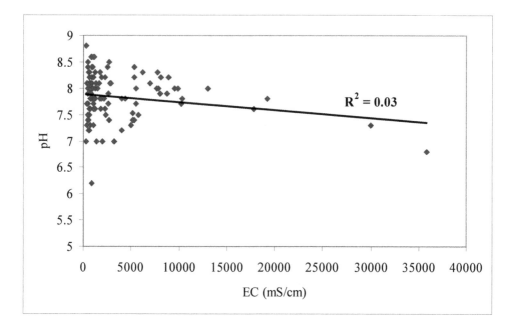

Fig. 3. Relationship between pH and EC values of the groundwater samples

7.2 Hydrogeochemical facies and classification

The chemical composition of groundwater is primarily dependent on the geology as well as on the geochemical processes which take place within the groundwater system. The Piper trilinear diagram (Piper, 1944) has been used for the purpose of characterizing the water types present in the area (Figure 4). Water types are often used in the characterization of waters as a diagnostic tool (Leybourne et al., 1998; Pitkanen et al., 2002). In addition, Piper diagram also permits the cation and anion compositions of many samples to be represented on a single graph in which major groupings or trends in the data can be discerned visually (Freeze and Cherry, 1979). Furthermore, it is used to assess the hydrogeochemical facies. A few conclusions can be inferred from the piper diagram of the collected samples (Figure 4). First, it shows that groundwater in the watershed is of very different types which evolve from bicarbonate to chloride type. Secondly, large percentages of the samples fall within the Na-Cl category followed by Na-HCO$_3$ type. Thirdly, chloride is the dominant anion found in the groundwater in the study area with the concentration ranging from 0.3 to 285 meq/L.

High concentration of Cl⁻ may be due to leaching of saline soil residues into the groundwater system, a typical characteristic of arid and semi-arid regions (Zaheeruddin and Khurshid, 2004). Finally, Piper plot shows that sodium with a mean value of 21.4 meq/L dominates the cationic components of the groundwater, although the sum of calcium and magnesium is higher that sodium. Figure 5 illustrates the spatial distribution of different water types throughout the watershed and shows that Na-Cl type waters are mainly concentrated in the plain part of the study area.

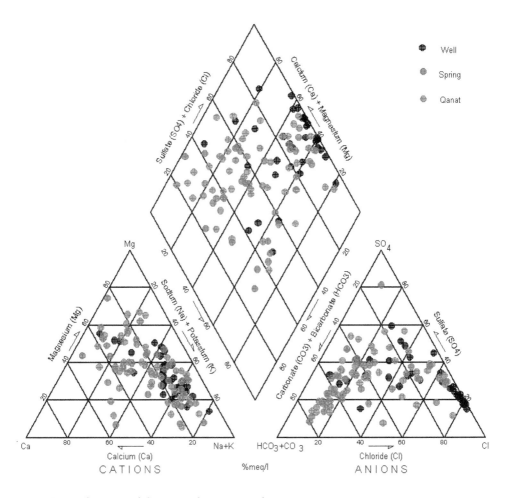

Fig. 4. Piper diagram of the groundwater samples

7.3 Origin of major ions

Concentration of ions dissolved in groundwater is generally controlled by lithology, groundwater flow rate, natural geochemical reactions and human activities (Karanth, 1997;

Bhatt and Salakani, 1996). In fresh groundwaters, both HCO_3^- and CO_3^{2-} originate mainly from the atmosphere but dissolution of sulfates, dolomite, calcite or silicates minerals also contribute to the concentration of these ions. Figure 6a displays the concentration of Ca^{2+} $+Mg^{2+}$ versus HCO_3^-. It is seen that all samples lie above the 1:1 line; this demonstrates an extra source for both Ca^{2+} and Mg^{2+} ions. The excess concentration of Ca^{2+} $+Mg^{2+}$ over HCO_3^- has been balanced by Cl^- and SO_4^{2-}. Excess calcium and magnesium is most likely supplied by the dissolution of various minerals such as dolomite, gypsum, calcite, anhydrite or weathering of silicate minerals such as plagioclase, pyroxene, amphibolites and montmorillonite (Freeze and Cherry, 1979, Boghici and Van Broekhoven, 2001). Plot of Ca^{2+} $+Mg^{2+}$ versus SO_4^{2-} $+Cl^-$ shows that all samples lie below theoritical 1:1 line, indicating the equilibrium between $SO_4^{2-}+Cl^-$ with alkalies (Figure 6b). Also plot of Ca^{2+} $+Mg^{2+}$ versus total cations (TC) illustrates that all samples lie far below the theoretical line depicting a contribution of alkalis to the major ions (Figure 6c). Groundwater in the area has a higher average ratio of Na^+ $+K^+$ versus total cations (Figure 6d). Plotting Na^+ $+K^+$ versus SO_4^{2-} show that all samples lie above 1:1 line (Figure 6e). The excess amount of Na^+ $+K^+$ over Cl^- (Figure 6f) at low chloride concentrations reflects input from the weathering of Na and K-rich minerals (Stallard and Edmond, 1983).

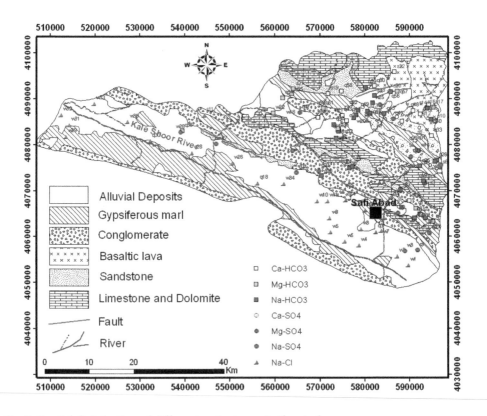

Fig. 5. Spatial distribution of different water types in the study area

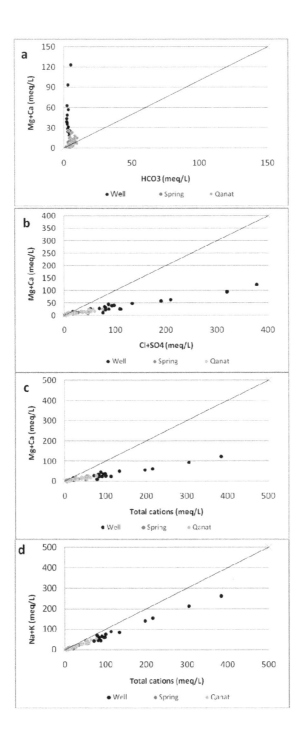

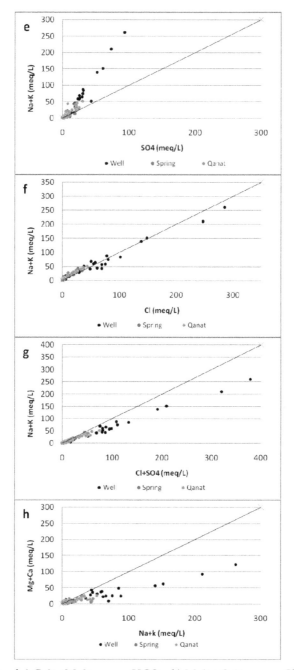

Fig. 6. Scatter plots of a) $Ca^{2+}+Mg^{2+}$ versus HCO_3^-, b) $Mg^{2+}+Ca^{2+}$ versus $Cl^-+SO_4^{2-}$, c) Ca^{2+} $+Mg^{2+}$ versus total cations, d) Na^++K^+ versus total cations, e) Na^++K^+ versus SO_4^{2-} f) Na^+ $+K^+$ versus Cl^- g) Na^++K^+ versus $Cl^- + SO_4^2$ and h) $Ca^{2+}+Mg^{2+}$ versus Na^++K^+.

Scatter plot of $Na^+ +K^+$ versus $SO_4^{2-} +Cl^-$ reveals that the increase in alkalis corresponds to a simultaneous increase in $Cl^- +SO_4^{2-}$ (Figure 6g), thereby indicating a common source for these ions and also the presence of Na_2SO_4 and K_2SO_4 in the soils (Datta and Tayagi, 1996). In samples with low molar $Na^+ +K^+$, $Ca^{2+} +Mg^{2+}$ and $Na^+ +K^+$ are balanced together and lie at 1:1 line with increase in molar values of $Na^+ + K^+$, data move to under the line (Figure 6h).

7.4 Factors governing water chemistry

The mechanism controlling water chemistry and the functional sources of dissolved ions can be assessed by plotting the ratios of Na^+ to $(Na^+ + Ca^{2+})$ and Cl^- to $(Cl^- + HCO_3^-)$ as functions of TDS (Gibbs, 1970). Gibbs diagram of the water samples (Figure 7) clearly shows that the majority of samples have become saline either via saline water intrusion or evaporative enrichment. As noted by some researchers, evaporation greatly increases the concentration of ions formed by chemical weathering, leading to higher salinity (Jalali, 2007). However, TDS values of over 1000 mg/L are mostly controlled by saline water intrusion. Figure 7 also shows that weathering of rock-forming minerals is only contributing marginally to the salinity of a small number of samples.

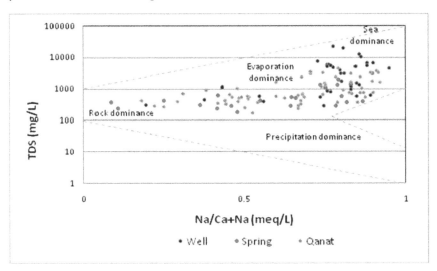

Fig. 7. Gibbs diagram of water samples

7.5 Geochemical modeling

Geochemical interactions often lead to changes in the water chemistry. The potential for a chemical reaction to happen can be determined by calculating the chemical equilibrium of the water with the mineral phases in question. The equilibrium state of the water with respect to a mineral phase can be determined by calculating saturation index (SI) using analytical data. Variation in the groundwater chemistries is mainly a function of the interaction between the groundwater and the mineral composition of the aquifer materials through which it moves (Nwankwoala and Udom, 2011). These processes generally include chemical weathering of minerals, sorption, and desorption, dissolution–precipitation of secondary carbonates and ion exchange between water and clay minerals (Apodaca et al.

2002). During rock weathering, Ca^{2+}, Mg^{2+}, SO_4^{2-}, HCO_3^- and SiO_2 are added to the water. The amount of each depends on the specific mineralogy of the rocks in question (Hounslow, 1995). In this research, the geochemical modeling program PHREEQC (Parkhurst and Appelo, 1999) has been employed to evaluate the water chemistry. The inverse modeling in PHREEQC takes into account uncertainty limits that are constrained to satisfy the mole balance for each element and valance state, as well as the charge balance for each solution within the simulation. In this research, we have employed PHREEQC to calculate saturation indices (SI) of sampled waters, only; it has many other applications.

Figure 8 shows the plot of SI of various minerals against TDS for all samples. With the exception of a few, all samples are super saturated with respect to both calcite and dolomite, suggesting that these carbonate mineral phases are extensively present in the corresponding host rock. Sharif et al., (2008) argue that the common ion effect of gypsum dissolution and calcite precipitation is often accompanied by dolomite dissolution leading to the meaningful increase in the magnesium content of groundwater. This is the case in the study area where higher numbers of samples are super saturated with respect to dolomite than calcite. All samples are under saturated with respect to anhydrite, gypsum and halite, even for very saline samples. Therefore, one may conclude that the interaction between groundwater and aquifer matrix is not significant in controlling chemical characteristics of groundwater in the study area, i.e. the source of the ions is mostly outside of the aquifer matrix.

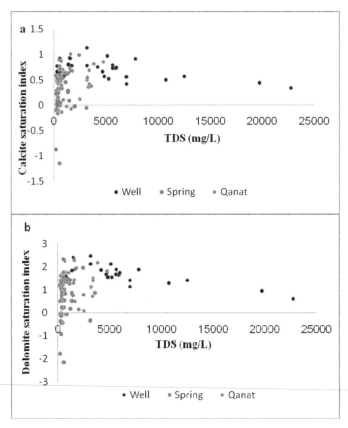

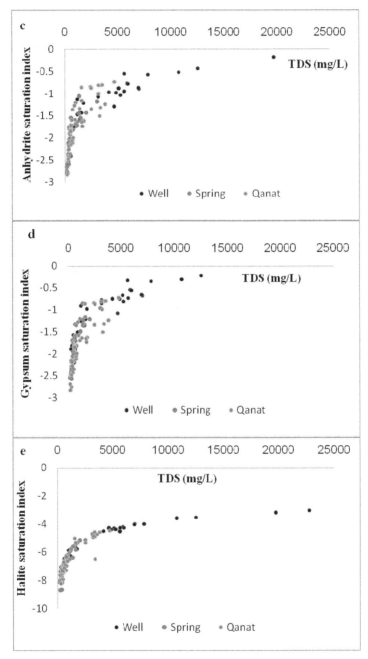

Fig. 8. Plot of the PHREEQC calculated saturation indices (SI) for calcite, dolomite, halite, gypsum and anhydrite versus TDS.

7.6 Ion exchange and reverse ion exchange

Ion exchange involves the replacement of ions adsorbed on the surface of the fine-grained materials of the aquifers by ions that are in the solutions (Todd, 1980). To investigate the importance of ion-exchange processes in groundwater chemistry, we have examined the relationship between the concentration of $(Na^+ - Cl^-)$ against $(Ca^{2+} + Mg^{2+} - SO_4^{2-} - HCO_3^-)$ as suggested by Boghici and Broekhoven (2001) and Jalali (2007). This is shown in Figure 9a. The product of $(Na^+ - Cl^-)$ represents excess sodium, that is, sodium coming from the sources other than halite dissolution, assuming that all chloride is derived from halite. Also, the product of $(Ca^{2+} + Mg^{2+} - SO_4^{2-} - HCO_3^-)$ represents the calcium and/or magnesium coming from sources other than gypsum and carbonate dissolution. In the absence of these reactions, all data should plot close to the origin (McLean et al., 2000). If these processes are significant composition governing processes, the relation between these two parameters should be linear with a slope of -1 (Figure 9a)

The plot of Na/Cl against EC is shown in Figure 9b. This shows that at higher ECs, the ration of Na/Cl decreases indicating that a) the ratio is approaching seawater ratio of 0.56 and b) reverse ion exchange is far more widespread than ion exchange. Figure 9c shows the amount of $Ca^{2+} + Mg^{2+}$ gained or lost relative to that provided by calcite, dolomite and gypsum. When $HCO_3^- + SO_4^{2-}$ is low (less than 5 meq/L) and the samples plot on 1:1 line, dissolution of calcite and dolomite is the major process influencing water chemistry but when $HCO_3^- + SO_4^{2-}$ is more than 5 meq/L, in addition to calcite and dolomite, dissolution of gypsum is also likely (Kalantary et al. 2007). For all samples of the Safi Abad aquifer, $HCO_3^- + SO_4^{2-}$ is >5 meq/L, indicating that gypsum dissolution is likely to occur. Almost all samples lie close to 1:1 line. However, some samples lie above the equiline, pointing to the ion exchange reactions. Also, many samples lie below the equiline indicating reverse ion exchange.

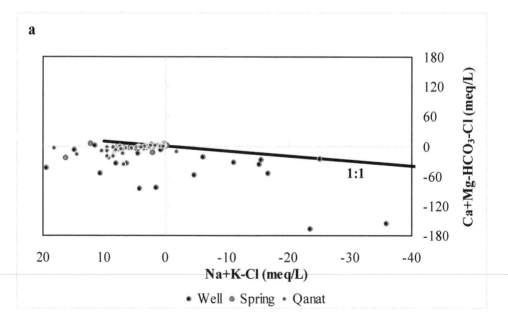

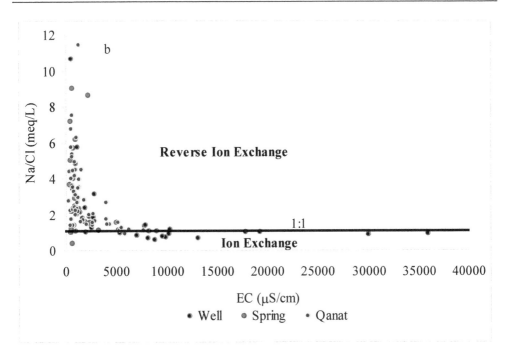

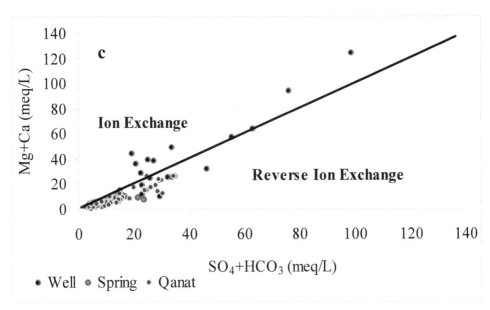

Fig. 9. Plot of a) $K^+ + Na^+ - Cl^-$ versus $Ca^{2+} + Mg^{2+} - SO_4^{2-} - HCO_3^-$, b) Na/Cl versus EC and c) $HCO_3^- + SO_4^{2-}$ versus $Mg^{2+} + Ca^{2+}$

8. Implication of hydrogeochemical studies for the management of local groundwater resources

To manage a groundwater resource, we should not only concern about the water quantity and what happens to it, but we should also take care of the quality. Deterioration of groundwater quality may result in serious restrictions on its usages especially when it is used for domestic purposes. The hydrogeochemical approaches implemented in this research shows that a number of groundwater wells and springs have turned into saline most probably due to encroachment of very saline water into these aquifers. Due to aridity, deep saline fossil waters are found in the vicinity of Iranian inland aquifers (e.g. Kazemi, 2011; Kazemi et al., 2001). Therefore, it is well possible that overexploitation of the fresh groundwater resources have impacted the saltwater-freshwater zone, leading to the movement of the fossil waters. If this is the case, this phenomenon will be intensified in the future due to increasing demand for water as well as global warming.

9. Conclusions

The quality of groundwater resources is as important as their quantity. Hydrogeochemical studies are a useful tool which can help managing quality of these resources. This research shows that rainfall percolating into the mountainous areas of Safi Abad watershed in the northeastern Iran remains mostly fresh while moving in the subsurface environment in the same mountainous area. However, the quality of such groundwaters deteriorates rapidly as it discharges into the adjoining plain. Different hydrogeochemical techniques show that a variety of causes lead to severe salinization of groundwater in the plain including saline water intrusion, leaching of local saline soils into underlying groundwater, intense evaporation and slow rate of groundwater movement. Over exploitation of the groundwater resources during the past 2 decades has probably changed the saltwater-freshwater interface. As a management scenario, the resource managers should first further explore the validity of this hypothesis. Also, it seems that withdrawal of fresh groundwaters in highland areas before letting it to discharge into the aquifer underlying the Safi Abad plain is a possible management scenario to avoid further salinization of this valuable resource.

10. References

Afshar Harb, A., Bolourchi, M. and Mehr Parto, M., 1987. Geological quadrangle map of Iran no. J5 (Bojnourd sheet), Scale 1:250,000, Geological Survey of Iran.

Apodaca, L.E., Jeffrey, B.B. and Michelle, C.S., 2002. Water quality in shallow alluvial aquifers, Upper Colorado River Basin, Colorado. Journal of American Water Research Association, 38(1):133-143.

Appelo, C.A.J. and Postma, D., 2005. Geochemistry, groundwater and pollution, Second Edition. Balkema, Leiden, The Netherlands, 683 p.

Atwia, M.G., Hassan, A.A. and Ibrahim, A., 1997. Hydrogeology, log analysis and hydrochemistry of unconsolidated aquifers south of El-Sadat city, Egypt. J. Hydrol., 5:27–38.

Back, W., 1966. Hydrochemical facies and groundwater flow patterns in northern part of Atlantic Coastal Plain. US Geol. Survey Professional Paper 498-A.

Ballukraya, P.N. and Ravi, R., 1999. Characterization of groundwater in the unconfined aquifers of Chennai City, India; Part I: Hydrogeochemistry. J. Geol. Soc. India, 54:1–11.

Bhatt, K.B. and Salakani, S., 1996. Hydrogeochemistry of the upper Ganges River. J. Geol. Soc. India, 48:171–182.

Boghici, R., and Van Broekhoven, G. A., 2001. Hydrogeology of the Rustler Aquifer, Trans Pecos Texas; in Aquifers of West Texas. Texas Water Development Board Report, 356: 207-225.

Datta, P.S. and Tayagi, S.K., 1996. Major ion chemistry of groundwater in Delhi area: chemical weathering processes and groundwater flow regime. J. Geol. Soc. India, 47:179–188.

Eagelson, P.S., 1979. The annual water balance. Journal of Hydraulic Division of ASCE, 105:923–941.

Eagleson, P.S., 1978. Climate, soil and vegetation. Water Resources Research, 14:705–776.

Fetter, C. W., 1998. Contaminant Hydrogeology, Second Edition. Prentice-Hall, Upper Saddle River, NJ, 500 p.

Freeze, R.A. and Cherry, J.A., 1979. Ground Water. Prentice-Hall, Englewood Cliffs, NJ, 553 p.

Gibbs, R.J., 1970. Mechanisms controlling world water chemistry. J. Sci., 17:1088-1090.

Hem, J. D., 1989. Study and interpretation of the chemical characteristics of natural water, Third Edition. U.S. Geological Survey Water-Supply Paper 2254, 263 p.

Hounslow, W.A., 1995. Water Quality data: Analysis and interpretation. Lewis Publishers, New York. 397 p.

Jalali, M., 2007. Salinization of groundwater in arid and semi-arid zones: an example Tajarak, western Iran. Environ Geol., 52:1133-1149.

Jalali, M., 2005. Major ion chemistry of groundwaters in the Bahar area, Hamadan, western Iran. Environ Geol., 47: 763–772.

Kalantary, N., Rahimi, M. and Charchi, A., 2007. Use of composite diagram, factor analyses and saturation index for quantification of Zaviercherry and Kheran plain groundwaters. J. Engg Geol. 2(1):339–356.

Karanth, K.B., 1997. Groundwater assessment, development and management. McGraw-Hill Publishers, New Delhi.

Kazemi, G. A., 2011. Impacts of urbanization on groundwater resources in Shahrood, Northeastern Iran: Comparison with other Iranian and Asian cities. Physics and Chemistry of the Earth, 36:150–159.

Kazemi, G. A., Fardoost, F., and Karami, G. H., 2001. Hydrogeology and groundwater quality of Shahrood Region, Iran. In: Proceed. of XXXI IAH Congress " New Approaches Characterizing Groundwater Flow" Seiler, K. P., and Wohnlich, S. (ed), Swets and Zeitlinger Lisse, Munich, Germany,1213-1216.

Leybourne, M.I., Goodfellow W.D. and Boyle, D.R., 1998. Hydrogeochemical, isotopic and rare earth element evidence for contrasting water-rock interactions at two undisturbed Zn-Pb massive sulphide deposits, Bathurst Mining Camp, N.B., Canada. J. Geochem Exp., 64:237-261.

McLean, W., Jankowski, J. Lavitt, N., 2000. Groundwater quality and sustainability in an alluvial aquifer, Australia. In: Sililo O et al (eds) Groundwater, past achievements and future challenges. A Balkema, Rotterdam, 567–573.

Merkel, B.J. and Planer-Friedrich, B., 2005. Groundwater geochemistry-A practical guide to modeling of natural and contaminated aquatic systems. Springer Verlag, Berlin, 200 p.

Nwankwoala, H.O., Udom, G.J., 2011. Hydrochemical facies and ionic ratios of groundwater in Port Harcourt, southern Nigeria. Research Journal of Chemical Sciences, 1(3):87-101.

Panigrahy P.I.C., Sahu S.D., Sahu B. K. and Sathyanarayana, D., 1996. Studies on the distribution of calcium and magnesium in Visakhapatnam harbour waters, Bay of Bengal. International Symposium on Applied Geochemistry, Osmania University, Hyderabad, 353–340.

Parkhurst, D. L. and Appelo, C. A. J., 1999. User's guide to PHREEQC (version 2): a computer program for speciation, batch reaction, one-dimentional transport and inverse calculation. US Geological Survey Water-Resources Investigations Report 99-4259, US Geological Survey, Reston, Virginia.

Piper, A.M., 1994. Graphical interpretation of water analysis, Transactions of the American Geophysical Union. 25:914 -928.

Pitkanen, P., Kaija J., Blomqvist R., SmellieJ.A.T., Frape S.K., Laaksoharju M., Negral P.H., Casanova J. and Karhu J., 2002. Hydrogeochemical interpretation of groundwater at Palmottu, Paper EUR 19118 EN, European Commission, Brussels.

Prasanna, M.V., Chidambaram, S., Shahul Hameed, A. and Srinivasamoorthy, K., 2011. Hydrogeochemical analysis and evaluation of groundwater quality in the Gadilam river basin, Tamil Nadu, India. J. Earth Syst. Sci. 120(1):85–98.

Ramappa, R., and Suresh, T.S., 2000. Quality of groundwater in relation to agricultural practices in Lokapavani river basin, Karnataka, India. Proceedings of International Seminar on Applied Hydrogeochemistry, Annamalai University, 136–142.

Sharif, M.U., Davis, R.K., Steele, K.F., Kim, B., Kresse, T.M. and Fazio, J.A., 2008. Inverse geochemical modeling of groundwater evolution with emphasis on arsenic in the Mississippi River Valley alluvial aquifer, Arkansas (USA). J. Hydrol., 350:41-55.

Stallard, R. F. and Edmond, J.M., 1983. Geochemistry of the Amazon, 2, The influence of geology and weathering environment on dissolved load. J. Geophys. Res., 88:9671–9688.

Stocklin, J., 1968. Structural history and tectonics of Iran: A review. Am. Asso. Petr. Geol. Bull., 52(7):1229–1258.

Subyani, A. M., 2005. Hydrochemical identification and salinity problem of groundwater in Wadi Yalamlam basin, Western Saudi Arabia. Journal of Arid Environments, 60:53–66.

Todd, D. K., 1980. Groundwater hydrology, Second Edition. John Wiley and Sons, New York, 535 p.

Tyagi, S.K., Datta, P.S. and Pruthi, N.K., 2009. Hydrochemical appraisal of groundwater and its suitability in the intensive agricultural of Muzaffarnagar district, Utter Pradesh, India. Environ Geol. 56:901-912.

Zaheeruddin and Khurshid, S., 2004. Aquifer geometry and hydrochemical framework of the shallow alluvial aquifers in the western part of the Yamuna River Basin, India. Water Qual. Res. J., 39(2):129–139.

4

A Review of Approaches for Measuring Soil Hydraulic Properties and Assessing the Impacts of Spatial Dependence on the Results

Vincenzo Comegna[1], Antonio Coppola[1],
Angelo Basile[2] and Alessandro Comegna[1]
[1]Dept. For Agricultural and Forestry Systems Management,
Hydraulics Division, University of Basilicata, Potenza
[2]Institute for Mediterranean Agro-Foresty, (ISAFOM),
National Research Centre, Ercolano
Italy

1. Introduction

The movement of water in the soil and associated solute transport perform a role of primary importance in many applications in the field of hydrology and agriculture. In the sound management of irrigation water, in relation to specific environmental conditions and cropping systems, knowledge of local water flow conditions in zones affected by the root systems is indispensable. Once the irrigation method has been established, only knowledge of the laws governing water flow allows the necessary irrigation frequencies and rates to be established to optimise the distribution of soil moisture, reducing within established limits the effects of water stress and containing water wastage.

Only by studying water dynamics in soil can the contribution of groundwater to water consumption be quantitatively determined. Moreover, the water volumes infiltrating into the soil due to rainfall are strictly linked and governed by the laws of water flow in the soil. No evaluation of water quantities being added to groundwater circulation can be made without first determining the water volumes moving in the zone between the soil surface and the aquifer.

Knowledge of water fluxes, velocities and contents is also indispensable to study the flow of solutes and pollutants and to predict all exchanges, whether chemical or microbiological, occurring in the soil. Within environmental protection initiatives, great weight is attributed to the continuous contribution of solutes characterising any agricultural activity, especially if intensive. Nor should we underestimate the hazards of non-agricultural solute fluxes, and the need to dispose of wastewater, muds and industrial effluent.

Except for situations of direct hazardousness, in general soil and groundwater degradation is linked to the different mobility that solutes may show in the various soil zones. Particular conditions may lead to accumulation zones for certain substances, while in other zones solutes may be easily transported in depth, with consequent alteration of local aquifer characteristics. To evaluate the nature of the risk represented the presence of these solutes it is important to define the processes governing their movement downward from the soil surface through the vadose zone as far as the aquifer. Only if we know such transport processes can optimal management plans be developed for environmental control with a view to preventing degradation phenomena.

The complexity of such flow processes has encouraged the widespread use of mathematical models corresponding as closely as possible to real phenomena which can supply quantitative evaluations also in the presence of highly complex systems, such as soil-plant-atmosphere continuum (SPAC). To be applied, such models, based on the laws of water flow in the vadose zone, require the relationships linking the volumetric water content, θ, the water potential h and the hydraulic conductivity, K.

Moreover, for field applications, an evaluation in statistical terms of the variability of θ-h-K relationships is also necessary.

Numerous researches have proposed testing techniques for laboratory and field soil characterization which, however, are almost always laborious and time-consuming (Klute, 1986; Hillel et al., 1998).

The choice of which methodology to adopt for these determinations depends in each case on the particular soil-type and on the local soil conditions. For this reason, it is preferable to have a preliminary soil profile pedological analysis revealing the existence of a distinct horizontal structure or of layers in the soil profile is of particular importance when large areas have to be characterized in terms of transport behavior (Bouma 1983; Coppola et al., 2011).

The purpose of this chapter is to present some experimental procedures for determining the unsaturated hydraulic properties both in the laboratory and open field. It was also developed in view to estimate the structure of the spatial variability exhibited by the available hydrological data set.

2. Soil water flow theory

Soil complexity and the presence of often interacting phenomena make it difficult to study particular aspects of specific processes that always have to be viewed in terms of overall system evolution. Thus in the sector of soil hydrology increasingly sophisticated mathematical models based on real phenomena are being applied. With the support of increasingly available numerical calculators, in many cases such models allow quantitative evaluations to be attained also in the presence of complex systems which are difficult to study (Jury et al., 1991).

In general, such models are based on laws governing water flow and all the physical and chemical processes affecting water transport. These laws have been widely confirmed by experiments and are often reported also in papers in disciplines collateral to soil hydrology (Bear, 1979; Jensen, 1980). Water flow is studied with reference to a porous medium which, on a macroscopic scale, may be considered continuous and in which the various quantities and physical properties are considered functions of position and time. Reference is generally made to isothermal flow processes, to a Newtonian liquid phase and interconnected gaseous phase with a pressure equal to that of the atmosphere. Moreover, due to relatively low

A Review of Approaches for Measuring Soil Hydraulic Properties and Assessing the Impacts
of Spatial Dependence on the Results

99

resistance to air flow, the movement of the gaseous phase is neglected and only water flow is referred to.

Such a schematization leads to concrete results when based on a generalized Darcy's law which in the case of isotropic porous media is expressed as follows:

$$v = -K \nabla H \tag{1}$$

in which v is the filtration velocity, K the hydraulic conductivity of the porous medium, ∇ is the Laplacian operator and $H = z + h$ the hydraulic potential of the flow domain, where z is the gravitational head and h the pressure head.

Unlike what happens in filtration processes through saturated porous media, the presence of interconnected air in the pores reduces the effective section of water flow, increasing path tortuosity, which is why conductivity K is variable with water content in volume θ. As regards the pressure head h, it should be pointed out that in unsaturated zones it assumes negative values (matric head), hence the widespread custom in the soil physics literature of referring to it as *suction* or *tension*. Further, as shown by laboratory and field experiments, the link between θ and h is not unique, but is characterised by a multi-value hysteretic function (Mualem and Dagan, 1975).

To study water flow, Darcy's law is coupled to the continuity equation which may be expressed as:

$$\frac{\partial(\rho\theta)}{\partial t} = -\nabla (\rho v) \tag{2}$$

in which ρ, as is customary, is used to indicate water density and t the time.

Allowing for Darcy's law we obtain from (2):

$$\frac{\partial(\rho\theta)}{\partial t} = \nabla (\rho K \nabla H) \tag{3}$$

The low water compressibility means that the variation of ρ can be neglected, hence (3) can be written as (Richards' equation):

$$C\frac{\partial h}{\partial t} = \nabla \left[K \nabla (h+z) \right] \tag{4}$$

in which $C = d\theta/dh$ is the differential water capacity of the porous medium. Or, introducing the diffusion coefficient $D = K dh/d\theta$, from (3) we obtain the Fokker-Plank equation:

$$\frac{\partial\theta}{\partial t} = \nabla (K \nabla z + D \nabla \theta) \tag{5}$$

Having established the initial conditions and those at the boundary of the flow domain, we can then integrate 4 or 5 only if we know at all points the functions $K(\theta)$ and $\theta(h)$ that completely characterise the porous medium in hydraulic terms. Moreover, due to the non-linearity of differential equations 4 and 5 and the fact that functions $K(\theta)$ and $\theta(h)$ (see

Appendix A) cannot always be given a satisfying analytical expression make it necessary adopt numerical integration techniques.

3. Methods of measuring soil hydraulic parameters

The widespread use of increasingly advanced numerical calculators gives us in practice unlimited possibilities of solving differential equations 4 and 5. However, so far there have been limited applications of these solutions to problems of practical interest despite the fact that in many situations there is a considerable need to achieve optimal management, with consequent savings, of irrigation water, as well as a need for soil conservation and groundwater resource protection.

The reasons for this slow spread of soil water flow theory may be sought in the difficult and burdensome task of soil hydraulic characterisation. In particular, hydraulic conductivity $K(\theta)$ is determined with laborious time-consuming methods, it undergoes high variations even with slight changes in water content and in many cases depends on the composition and concentration of circulating solution in the soil (Frenkel et al., 1978). Moreover, the functions $K(\theta)$ and $\theta(h)$, especially at higher water contents, differ according to whether there is a wetting or drainage phase. The choice of methodologies to use for measuring hydraulic characteristics depends on the particular soil and conditions at the measuring point, which is why it is preferable that surveys are preceded by a pedological study. Especially in the case of clayey soils, a critical analysis is required concerning the applicability of the measurement method, the type of probes to be used and the dimensions of the soil volume to which measurements refer.

Hydraulic characteristics may be determined using field or laboratory experiments, operating on undisturbed samples. The validity of the results is subordinate to the condition that the field installation of neutron and/or Time Domain Reflectometry (TDR) probes (Hillel, 1998) or taking of samples does not disturb the porous medium (Basile et al., 2003, 2006). Both in field and laboratory experiments the system to be characterised usually concerns very simple one-dimensional flow processes of wetting or drainage. Appropriately positioned measuring devices allow, in time and at specified depths, water contents θ and h to be recorded. With simultaneous measurements of θ and h it is possible to obtain at various points in the flow domain the functions $\theta(h)$, while conductivity values are determined by means of Darcy's law, inferring from the evolution of θ and h profiles, the gradients of the potential and water flux density at various depths (Hillel, 1998).

In the following, some methodologies will be illustrated which are widely used for hydrological characterizations of unsaturated porous media.

3.1 The evaporation method

According to this method, the determination in the laboratory of the hydraulic properties of soil can be made by submitting undisturbed soil samples to a process of evaporation of the upper surface, having taken care to seal off the lower one in order to prevent water-loss from the bottom (Figure 1).

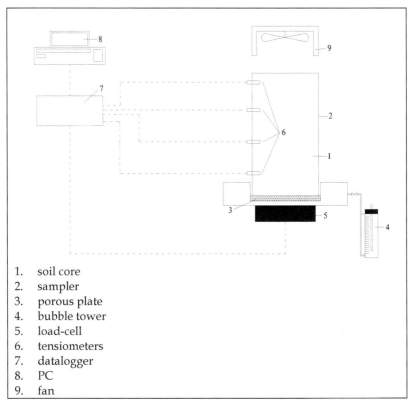

1. soil core
2. sampler
3. porous plate
4. bubble tower
5. load-cell
6. tensiometers
7. datalogger
8. PC
9. fan

Fig. 1. Scheme of experimental setup of the evaporation method

This method was originally developed by Wind (1968) and subsequently used by Boels (1978) and Tamari et al. (1993).

The soil samples, after depth saturation, can be dried under thermoregulated conditions employing a small fan with the same diameter as the sample and positioned about 10 cm above its upper surface.

Following the starting of the fan, a constant evaporation flow is rapidly determined, the value of which can be varied within certain limits by regulating the air-flow which strikes the upper surface of the sample.

During the transient state, simultaneous measurements of the weight P of the sample can be taken, using strain-gauge load cells with a measuring range from 0 to 6 kg and precision ± 1 g and of potential h to 4 or 5 depths z of the sample. The measurement of the potential values is effected by boring on the sampler wall a fixed number of holes, into which are then inserted miniaturized tensiometers with a porous septum of sinterized glass (10 mm diameter and 30 mm length, bubbling pressure at 500 hPa) joined to pressure transducers.

The measuring and recording of the data can be carried out by the data control and acquisition system, driven by a numerical calculator, and the depletion process is interrupted at the observation of the entrance of air into two tensiometers.

Following this, with the aim of extending the curve θ(h) to lower potential values, the sample is removed and, after having been broken down into sections of depth equal to 2 cm, is submitted to further draining on the membrane plate apparatus (Klute, 1986).

The measurements of evaporation flows and of potentials at different depths permitted the evaluation of both the retention curve θ(h) and the conductivity curve K(θ).

In fact, assuming that the sample can be characterized by a single equation:

$$\theta = f(h) \tag{6}$$

and indicating with ΔP_t, the weight of evaporated water from the beginning of the transient until time t we have:

$$\Delta P_t = A \int_0^L \gamma \left[\theta(z) - \theta_0(z) \right] dz \tag{7}$$

in which $\theta_0(z)$ indicates the initial distribution of water content, and A the cross-section of the sample.

Equation (7), then, taking into account equation (6), can be set out in the form:

$$\Delta P_t = A \int_0^L \gamma \left[f(h(z)) - f(h_0(z)) \right] dz \tag{8}$$

If finally, the sample is divided into n strata of range Δz ,and if we assume that the measured h as taken by each tensiometer correspond to the mean values of the strata, (equation 8) can be approximated by:

$$\Delta P_t \cong A\gamma \sum_i^n \left[f(h_{t,i}) - f(h_{0,i}) \right] \Delta z \tag{9}$$

in which $h_{t,i}$ indicates the potential h measured to time t by the tensiometer of stratum i.

Following the suggestions of Boels et al. (1978) the equation θ=f(h) can be approximated with a polygonal with k vertices for which, when the axes h_j^* of the vertices are fixed, the equation θ=f(h) will be sectionally linear and can be defined once the values θ_j^* of water content corresponding to potentials h_j^* are known.

Therefore, once the weight variation ΔP_t and the potentials $h_{t,i}$. and $h_{0,i}$ are known , and the values of h_j^* fixed, equation (9) can be reduced to a linear equation in the unknown quantities θ_j^* . The determination of the values θ_j^* is possible if the potentials and weights are experimentally estimated at different times, and if the system of linear equations, obtained by writing (9) at least for K values of t, is resolved.

The precision of these evaluations can be increased within certain limits by raising the number of equations and therefore of the number of measurements.

The choice of the values h_j^* will be arbitrary, but in each case, in order to obtain a uniform approximation of the tension curves through the entire range being surveyed it is necessary

A Review of Approaches for Measuring Soil Hydraulic Properties and Assessing the Impacts
of Spatial Dependence on the Results
103

to have the values $h_{t,i}$ when measured, fall uniformly into all the intervals between two values h_j^* .

To make practicable the evaluations of θ_j^* a code can be prepared, which is able automatically to fix the most convenient coordinates h_j^* of the vertices of the polygonal $\theta=f(h)$ and to resolve the system of equations (9) by minimizing the sum of the squares of the residuals.

Once the law $\theta(h)$ had been obtained using the foregoing procedure, and using the potential profile $h(z,t)$, it is possible to turn to the water content profiles $\theta(z,t)$ and to calculate the conductivity K in the function of θ, using the instantaneous profile method in the way demonstrated earlier (Hillel, 1998).

3.1.1 A case study

In this section, we refer to a study directed at the hydraulic characterization of a silty-loam soil of the southern Italy. The soil develops on a recent alluvial formation of material of marine and fluvial origin, containing sandy clay lying on gravel beds. With reference to the 7th approximation proposed by the Soil Survey Staff of the U.S. Department of Agriculture, it is an association in which predominate fluvents vertic and vertic soils, with the former represent most strongly.

As to the texture, it is made up of silty-loam soils (sand 25%, silt 51%, clay 24%) to a depth of ≈110 cm, changing to loam/sandy-loam (sand 60%, silt 28%, caly 12%) at depths over 110 cm.

Six measuring sites were individuated which, taking into account preliminary inquiries into the structure of spatial variability of the hydraulic characteristics, may certainly be considered spatially independent (Gajem et al. 1981).

On each site, by means of the digging of a pit to the depth 10-40 cm, 40-60 cm and 60-80 cm, samples of undisturbed soil of diameter 80 mm and height variable between 140 and 160 mm, were taken. For each sample, we determined texture, bulk density, hydraulic conductivity at saturation point, and curves $\theta(h)$ and $K(\theta)$.

The percentage of sand, silt and clay, in keeping with the scheme proposed by the ISSS, was determined using the densimetric method (Day, 1965). Bulk density was obtained by drying the soil samples in an oven at a temperature of 105° C, basing the calculation on soil weight and on the volume of the cylindrical container.

The determination of the hydraulic properties of the soil samples can be made using the evaporation method.

In Figure 2 is presented, by way of example, the time-flow values of the evaporation of water volume at the top of samples no.34 and no.40, taken from the depths $z=10-40$ cm and $z=60-80$ cm respectively.

The evaporation process is determined by external energy, combined with the capacity of the soil to make available to the surface of evaporation the relative volumes of water.

Therefore, in general the process evolves in two distinct phases, of which the first is characterized by constant flow which depends only on the difference in vapour concentration between the air and the soil surface and on the laminar boundary resistances of the air.

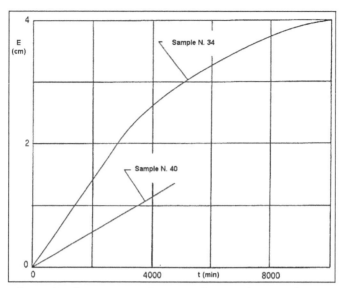

Fig. 2. Relation of cumulated evaporation E to time, for samples 34 and 40

When the soil surface reaches a level of water content such as not to permit the carriage of adequate water from the lower layers, the process becomes characterized by a reduction in time of evaporation flows, which is ever more marked by the reduction in water content.

These two phases are clearly visible in Figure 2, in which the evaporation process in sample no.34 was protracted for longer values of time.

From the graph, it can be seen that the flow of initial evaporation from sample no.40 is reduced. In fact it was necessary, in the case of samples characterized by low conductivity at saturation-point, to reduce the ventilation flow so as to avoid over-rapid depletion of the upper layers, which leads to potential profiles characterized by gradients which are too high at the surface and which cannot therefore be evaluated to an acceptable approximation.

However, in the tests it was necessary, depending on the particular characteristics of the samples, to regulate the flow of air so that water movement within the soil be sufficiently well distributed.

In Figures 3 and 4, for the two samples above mentioned, the potential values at different measuring depths are presented. From the graphs, we see that the potentials are clearly well differentiated from one another and that they evolve regularly in time, in such a way as to facilitate processing.

The results of the formulations carried out for all the samples according to the methodology illustrated allowed us to arrive at the presentations θ(h) of Figures 5 and 6, in which are shown the points specifying the vertices of the polygonals with which the retention curves were approximated.

Figure 5 in particular refers to samples taken from a depth of 10-40 cm and shows how the values of maximum water content are arranged in a band between 0.46 and 0.52, while for h equal to -250 cm, the values of water content are between 0.32 and 0.38.

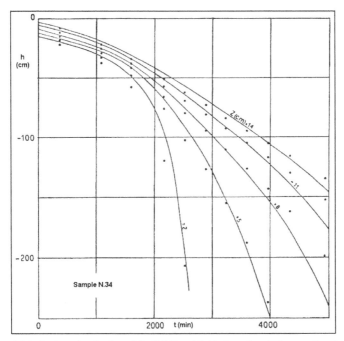

Fig. 3. Measured (circles) and calculated (solid lines) h(t) data for different depths of sample 34

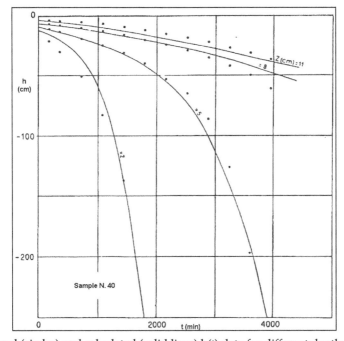

Fig. 4. Measured (circles) and calculated (solid lines) h(t) data for different depths of sample 40

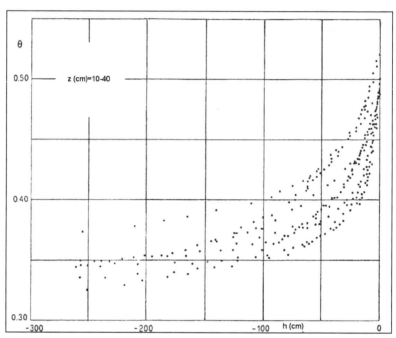

Fig. 5. Pooled soil water for all samples of selected depths 10-40 cm

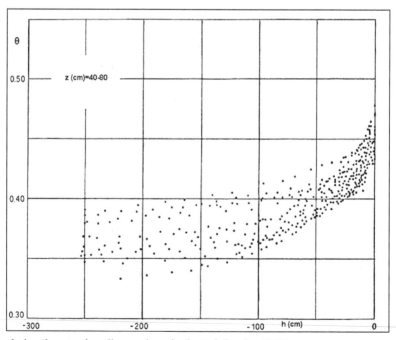

Fig. 6. Pooled soil water for all samples of selected depths 40-80 cm

Figure 6 refers to all remaining samples taken at a-depth of 40-80 cm, in so far that it was not possible to make distinctions between them. From the graph, it can be seen that the values of maximum water content are lower, and fall within a narrower band than those of Figure 5.

For a more practicable comparison, in Figure 7 are presented the mean retention curves for the two depth bands and the range of values of θ, into which it is expected that 68% of the water content will fall. The comparison shows how the mean maximum water content is equal to 0.50 for the surface samples and equal to a little over 0.45 for the deeper layer which, at potential values of -250 cm, is characterized, on average, by higher water content than the superficial layer.

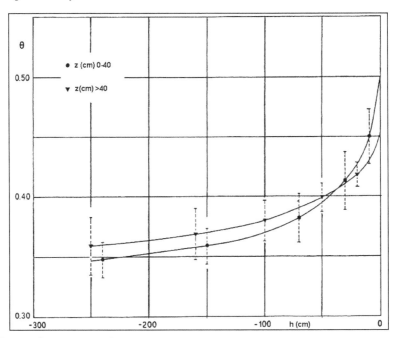

Fig. 7. Mean soil water retention curves

The hydraulic conductivities relative to the two ranges of depth, calculated using the methodology above described, are presented in Figures 8 and 9 in the function of water content. It can be seen that, as for all the samples, water conductivity is considerably reduced with a diminution of water content and that, despite there being a notable dispersion of points, slightly lower conductivities correspond to the greatest depths.

This test methodology was shown to be highly practicable, in that it did not call for the use of sophisticated equipment. The only evaluations necessary were of the weight of the samples and of the potentials at different depths.

To be able to test what the measurements mean and their precision, in view of one of their possible uses in simulation models, calculations were then carried out using the mathematical model, established in order to facilitate the study of evaporation processes in bare soil. The model was used to reconstruct the transient of evaporation effected during the tests, by attributing to each sample the hydraulic properties determined using the methodology previously illustrated.

With the aim of greater practicability, the analytic expression proposed by van Genuchten et al. (1980) was applied to the retention curve θ(h); while the conductivity curve was approximated with an exponential function at two parameters. The parameters shown in the above-mentioned equations were then determined by adapting, through the optimization method, the analytic curves to the experimental data.

The results of these calculations were shown to be quite similar to the measurements taken during the tests. In particular, continuing to refer to sample nos. 34 and 40, in Figures 3 and 4, together with the curves measured experimentally, are shown the results of the numerical reconstructions carried out. In these graphs it can be seen that the dispersion of the points is modest: this can, however, be attributed to the imperfect homogeneity of the samples.

Therefore the verification calculations we carried out show how the test method employed is sufficiently precise for application in mathematical models.

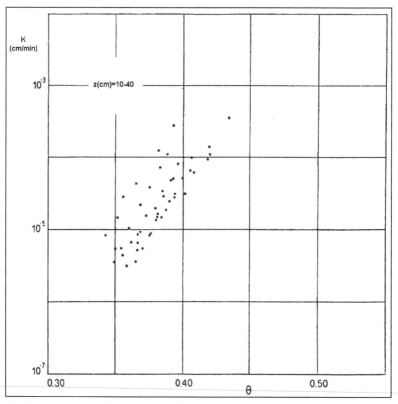

Fig. 8. Pooled hydraulic conductivity data for all samples of selected depths 10-40 cm

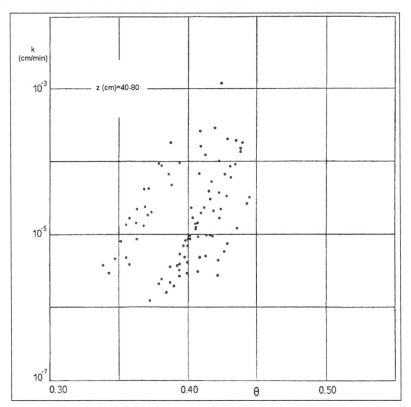

Fig. 9. Pooled hydraulic conductivity data for all samples of selected depths 40-80 cm

3.2 Internal drainage method

The water retention function $\theta(h)$ is generally determined directly in the laboratory by exposing undisturbed soil samples to wetting and drying cycles and carrying out measurements in equilibrium conditions or in the field by measuring several times concurrently the water content θ and the potential h (Klute et al. 1986).

It is far more difficult and laborious to determine hydraulic conductivity, in that it varies by several orders of magnitude, not only for different soils, but also for the same soil, as a function of water content. Moreover, for some soils laboratory methods are inadvisable: global characterisation is required with the use of direct measuring methods in the field.

For field determinations, during a drainage process, the liquid phase is almost always monitored by tensiometers which allow water potential h to be measured at different depths and by using TDR and/or neutron probes to measure the water content in volume θ. The data thus acquired are processed to define, for each measurement depth, the corresponding retention curve $\theta(h)$. Moreover, with the availability of profiles of potential h(z) and water content $\theta(z)$ the conductivity curve $K(\theta)$ can be directly obtained with the instantaneous profile method (Watson, 1966).

Such field tests are both time-consuming and burdensome, partly due to the difficulty with the automation of data collection. However, they provide representative results for the scale of measurements and, in some situations, are the only results achievable (Hillel, 1998).

3.2.1 Theory

Profiles of water content $\theta(z)$ and potential $h(z)$ allow us to determine hydraulic conductivity with the instantaneous profile method. Indeed, by integrating the differential equation of one-dimensional water flow in unsaturated media:

$$\frac{\partial \theta}{\partial t} = \frac{\partial}{\partial z}\left[K(\theta)\left(\frac{\partial h}{\partial z} - 1\right)\right] \tag{10}$$

between z=0 and depth Z and assuming that there is zero flow at the soil surface, we obtain:

$$\int_0^Z \frac{\partial \theta}{\partial t} dz = \left|K(\theta)\left(\frac{\partial h}{\partial z} - 1\right)\right|_{z=Z} \tag{11}$$

hence:

$$\left|K(\theta)\right|_{z=Z} = \frac{\int_0^Z \frac{\partial \theta}{\partial t} dz}{\left|\frac{\partial h}{\partial t} - 1\right|_{z=Z}} \tag{12}$$

Equation (12), expressed in discrete form, allows hydraulic conductivity to be determined by using experimental data. Indeed, with reference to a time interval $t_{i+1}-t_i$, and corresponding variations in water content $\theta_{i+1}-\theta_i$, we have as follows:

$$K(\bar{\theta}) = \frac{\int_0^Z (\theta_i - \theta_{i+1}) dz/(t_{i+1} - t_i)}{\left|\frac{\partial h}{\partial z} - 1\right|_{t=(t_{i+1}-t_i)/2}^{z=Z}} \tag{13}$$

where $\bar{\theta}$ may be determined using the relation:

$$\bar{\theta} = \frac{1}{2}\left(\theta_{i+1}(Z) - \theta_i(Z)\right) \tag{14}$$

and the derivative $\partial h / \partial z$ may be evaluated with finite differences from the measurements of potential h.

3.2.2 A case study

In this section we refer to a survey to characterize a volcanic topsoil hydraulically determining $K(\theta)$ and $\theta(h)$ based on the instantaneous profile method.

The experimental field soil, which is located at the Ponticelli-site (nearby Naples, Italy), is a sandy soil (sand 80%, silt 12%, clay 8%), pedologically classified as andosol. The main feature is that the soil is macroscopically homogeneous up to 0.80 m, with a layer of finer textured (loamy) soil at 0.80÷1.00 m. Measurements of the soil bulk density ρ up to 1.00 m (with 0.30 cm depth intervals) at various locations across the field had mean and standard deviation equal to 1.01 g/cm³ and 0.13 g/cm³), respectively.

For the purposes of soil hydraulic characterization, in terms of the water retention θ(h) and hydraulic conductivity curves K(θ), a 40x7m² plot was set up (Figure 10). On a side transect of the plot 120 three-wire steel probes were installed at different depths (z=0.30, z=0.60 and z=0.90 m), spaced 1 m apart, to measure water content with the time domain reflectometry method (TDR) (Topp et al., 1980). Along a parallel transect 0.50 m from the probe transect 120 tensiometers were installed to measure water potential h.

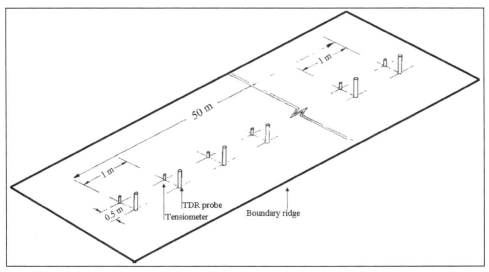

Fig. 10. View of the experimental plot displaying relative position of TDR probes and 0.3 m depth tensiometers.

As commonly used for field determination of soil hydraulic properties, when infiltration was completed, a drainage test was then carried out. With reference to the instantaneous profile method, potential h and water content θ are measured at the same time during a process of redistribution of the liquid phase, with evaporation and rainfall infiltration prevented at the soil surface by a plastic sheet.

Measurements of water content and potential were conducted in 12 campaigns and interrupted about 80 days from the end of infiltration, when the drainage process was evolving so slowly as to make it fruitless to continue measurements.

To increase the method's rapidity and versatility, water content and potential were measured by connecting the TDR probes and tensiometers to new-generation portable testers (Hopmans et al 1999; Carlos et al, 2002): TDR 100 testers (Campbell Scientific Inc, Logan UT) and a tensimeter-microdatalogger (Skye Instruments Ltd., UK). The calibration curve of the TDR probes was determined with the method proposed by Regalado et al. (2003) for volcanic soils.

At the end of the test, at each measuring site, at the average profile depth of 0.30 m soil hydraulic conductivity was measured at saturation K_s with the constant-head well permeameter technique (Amoozegar and Warrick, 1986). At the same depth, disturbed soil samples were taken to determine texture with the densimetric method (Day, 1965).

With reference to one of the profiles examined (no. 5) along the transect, Figures 11, 12 and 13 illustrate the extent of the transients induced in the porous medium during the test.

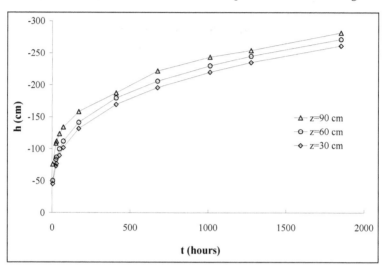

Fig. 11. Field values of h(t) for the drainage transient of profile 5

Water potential measurements (Figure 11) at various depths show that the potentials are sufficiently differentiated and evolve in time with regularity so as to permit straightforward calculation of the gradient.

Total hydraulic potential H distributions with depth z and for some test times are reported in Figure 12. The continuous lines indicate that the profiles obtained may be described with good approximation by a linear relation.

Finally, θ is plotted against time t (Figure 13). After the rapid fall in water content during the first few days of drainage, θ values continue to drop significantly and at the end of the test appreciable reductions in water content over time are found for all depths. This shows that the soil in question has a considerable natural drainage capacity and that it is in practice impossible to define correctly the so-called "field water capacity" of the topsoil.

The elaboration of the whole data pool allowed us to obtain the h(θ) and the K(θ) reported in Figure 14a and 14b, respectively. The scatter of θ(h) points in Figure 14a is contained in a fairly narrow bound and shows marked homogeneity of the soil in question. By contrast the K(θ) values in Figure 14b show greater dispersion, even if somewhat restricted compared to those from other studies (Jury et al., 1991).

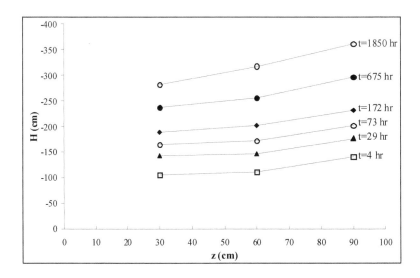

Fig. 12. Field values of H(z) for different time relative to the drainage transient of profile 5

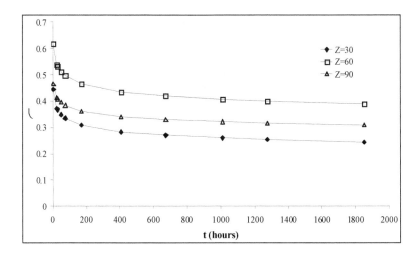

Fig. 13. Field values of θ(t) for the drainage transient of profile 5

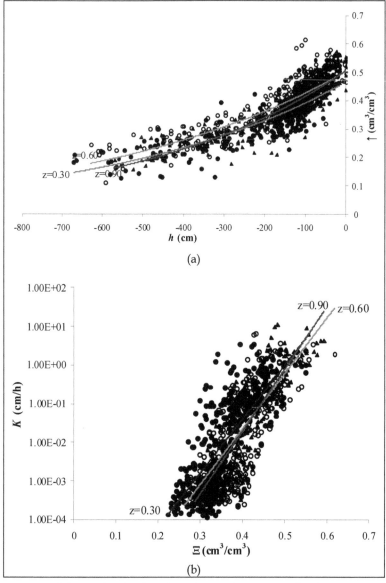

Fig. 14. a. Scatter of θ(h) and b. of K(θ) values detected at various depths of the soil profiles along the transect

3.3 Simplified analysis of internal drainage experiments

To make the measurements less laborious, for determining only hydraulic conductivity, simplified methods have been proposed (Libardi et al., 1980; Chong et al., 1981) whereby only one variable, θ or h, needs to be measured in the field, during the drainage process.

Hydraulic conductivity is evaluated by making simplifying hypotheses on the distribution of water potential gradients during the experiments and by assigning analytical expressions to relations $K(\theta)$ or $K(h)$ and to the laws by which θ or h vary in time.

The above expressions, generally with no more than 3 parameters, have been confirmed experimentally for very different soils, albeit with a homogeneous profile (Nielsen et al., 1973; Dane, 1980; Sisson and Van Genuchten, 1991), whereas they have been found inadequate in the case of soils characterized by successions of numerous strata with different physical characteristics (Schuh et al., 1984; Jones and Wagenet, 1984; Ahuja et al., 1988).

3.3.1 Theory

The methods of Libardi et al. (1980) and Chong et al. (1981) for calculating the conductivity relation are based on Richards' differential equation which, for one-dimensional vertical flow has the form:

$$\frac{\partial \theta}{\partial t} = \frac{\partial}{\partial z}\left[K(\theta)\frac{\partial H}{\partial z} \right] \tag{15}$$

Furthermore, the following simplifying assumptions were made:

1. The soil water flux at time t=0 is constant in the whole soil profile $(0 \leq z \leq L)$, and for $t > 0$ at the soil surface the zero flux condition occurs.

For these cases, by integrating equation (15) we obtain:

$$\int_{z=0}^{z=L} \frac{\partial \theta}{\partial t} = \left[K(\theta)\frac{\partial H}{\partial z} \right]_{z=L} \tag{16}$$

If θ^* represents the average water content of the profile, up to the maximum depth L, equation (16) is reduced to:

$$L\frac{d\theta^*}{dt} = \left[K(\theta)\frac{\partial H}{\partial z} \right]_{z=L} \tag{17}$$

2. The average relationship between θ^* and the water content $\theta(L,t)$, at depth is assumed to the form:

$$\theta^* = a\theta + b \tag{18}$$

where a [L] and b[L] are empirical constants.

3. Between the hydraulic conductivity k and water content θ, we assume the following exponential relation:

$$k(\theta) = k_0 \exp\left(\beta(\theta - \theta_0) \right) \tag{19}$$

where $\beta[L^3, L^{-3}]$ is a constant, and K_0 and θ_0 are the hydraulic conductivity and water content values at the beginning of the drainage process.

4. During the drainage process, the unit hydraulic head gradient is assumed to be at the greatest depth $z=L$:

$$\frac{\partial H}{\partial z} = -1 \tag{20}$$

With these assumptions, integrating equation (17) and by means of equations (18)-(20) the following relation is obtained:

$$-aL\frac{d\theta}{dt} = K_0 \exp\left(\beta(\theta - \theta_0)\right) \tag{21}$$

Where θ_0, K_0 and β apply at depth L.

To estimate K_0 and β the methods used are those proposed by Libardi et al. (1980) and Chong et al. (1981), namely: (1) θ method; (2) Flux method; (3) Lax-θ method; (4) CGA method (abbreviation for Chong-Green-Ahuja), which are briefly described below.

θ method

By integrating equation (21) with the initial conditions ($\theta=\theta_0$) we obtain for long times (t>2h):

$$\theta_0 - \theta = \frac{1}{\beta}K\ln(t) + \frac{1}{\beta}\ln\left(\frac{\beta \cdot K_0}{a \cdot L}\right) \tag{22}$$

Once a is known, the slope and the intercept with the axes of the graph of equation (22) allow parameters β and K_0 to be directly estimated.

Flux method

If the logarithms of the left-hand and right-hand members of equation (21) are considered, the following expression is obtained:

$$\ln\left|a \cdot L\frac{\partial\theta}{\partial t}\right| = -\beta(\theta_0 - \theta) + \ln(K_0) \tag{23}$$

With reference to the average values, which are more stable in the water content measurements, especially if carried out at greater depths, equation (23) may be also be expressed as:

$$\ln\left|a \cdot L\frac{\partial\theta^*}{\partial t}\right| = -\beta(\theta_0 - \theta) + \ln(K_0) \tag{24}$$

which describes a linear semi-log relationship where β and K_0 are determined from the slope and intercept, respectively.

Lax-θ method

The unit gradient hypothesis allows equation (15) to be written in the differential form:

A Review of Approaches for Measuring Soil Hydraulic Properties and Assessing the Impacts
of Spatial Dependence on the Results

117

$$\frac{\partial \theta}{\partial t} = -\frac{dK}{d\theta}\frac{\partial \theta}{\partial z} \tag{25}$$

the solutions to which are supplied by Lax (1972) and Sisson et al. (1980) for different analytical expressions for $K(\theta)$.

When $K(\theta)$ is given by equation (19), we obtain:

$$\theta_0 - \theta = \frac{1}{\beta}\ln(t) + \frac{1}{\beta}\ln\left(\frac{\beta \cdot k_0}{L}\right) \tag{26}$$

in which β and K_0 may be calculated from the slope and intercept, respectively, of $(\theta_0-\theta)$ versus $\ln(t)$. It should be noted that equation (26), does not require the assumptions of equation (18) and thus it is not necessary to evaluate factor a.

CGA method

If, instead of using equation (18), we assume that there is the following type of relation between θt^* and t:

$$\theta^* = A \cdot t^B \tag{27}$$

then we obtain, instead of equation (19), the expression:

$$K(\theta*) = -LA^{\frac{1}{B}}B\theta^{*(B-1)}\!\!\big/\!B \tag{28}$$

where coefficients A and B may be obtained by fitting equation (27) to experimental θ versus t data.

The β and K_0 values can be obtained from equation (28), if it is assumed that equation (18) holds and that the term $\left[a(\theta-\theta_0)\right]/\theta_0^*$ is small with respect to 1:

$$K_0 = -LA^{\frac{1}{B}}B\theta^{*(B-1)}\!\!\big/\!B \tag{29}$$

and

$$\beta = a\frac{B-1}{B\theta^*} \tag{30}$$

The estimates of A and B in equation (27) and of a in equation (18) thus allow us to evaluate parameters β and K_0 according to the other methods.

3.3.2 A case study

This section sets out to analyze the possibility of using limited data collected in site to determine the hydraulic properties of a volcanic soil by simplified methods proposed by Libardi et al. (1980) and Chong et al. (1981). The experimental site is located at Ponticelli (Naples, Italy). The soil profile is uniform with a moderate layering in the top metre and is classifiable as an Andosol. At the centre of the field where the trial was carried out, a plot with dimensions of 5 m x 5 m was prepared with a boundary ridge about 0.25 m high. At the centre of the plot two aluminium tubes were inserted up to a depth of 2.0 m, for measuring at depths: 30, 45, 60, 75, 105, 135 and 150 cm, water content by means of a RONLY 701A-11-P neutron probe. To reduce instrumental variance, which essentially depends upon the random variation in the fast-neutron emission from the source, three readings per depth were averaged. The error in the volumetric water content measurements were assumed to be due mainly to the calibration curve standard error of estimate equal to 1.7%.

At a distance of 0.5 m from tube axes, nine mercury-water manometer-type tensiometers were installed with their tip at a depth of 15, 30, 45, 60, 75, 90, 105, 120 and 150 cm to register water potential. The ceramic tensiometer cups were made in our laboratory, with the following characteristics: (i) the bubbling pressure (Pa) is greater than 500 hPa; (ii) the cup conductance (C) is greater than 1.11×10^{-5} cm^3 s^{-1} hPa^{-1} of pressure difference across the wall; (iii) considering that the gauge sensitivity (S) is 103 hPa cm^{-3}, an instrumental time constant in water $\tau = C^{-1}S^{-1}$ may be calculated equal to 90 s.

For the purposes of the trial, the plot was ponded by applying water in excess of the infiltration rate, while an overflow pipe guaranteed a constant water depth of 0.15 m. The time required for establishing steady-state flow in the profile at all depths to 1.5 m, was about 1 week. When infiltration was completed, the surface of the plot was covered with a plastic sheet so as to prevent evaporation from the soil surface and rainfall infiltration in the soil profile (see Figure 15).

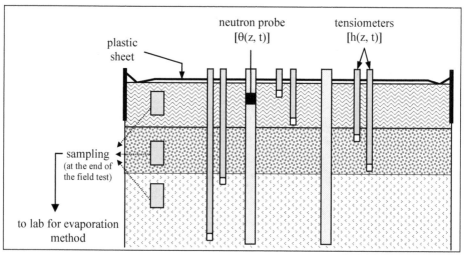

Fig. 15. Schematic of internal drainage setup

Measurements were carried out at increasing time intervals from the start of the drainage process. The water content θ and the potential h were always measured at the same time. Monitoring was interrupted 42 days after the end of infiltration when the drainage process was evolving so slowly as to make it pointless to continue with the experiment.

On completion of the trials at every measuring point in the profile, undisturbed soil samples
were taken to determine the texture by the densimetric method (Day, 1965).

Table 1 reports the main soil physical properties measured at various depths.

Depth z (cm)	Coarse sand (2<d<0.2 mm)	Fine sand (0.2<d<0.02 mm)	Silt (0.02<d<0.002 mm)	Clay (d<0.002 mm)	Bulk density (g)cm³)
10-20	30	50	12	8	1.15
40-50	30	53	12	5	1.10
70-80	34	48	12	6	1.09
100-110	25	63	7	5	1.09
110-120	27	60	5	8	1.09

Table 1. Texture and bulk density of the examined profile

Figure 16 shows the distribution $h(z)$ of with respect to time. The $h(z)$ curves have a regular
pattern for all depths and $\partial H/\partial z \approx 0$ near the surface where evaporation was prevented by
the plastic sheeting.

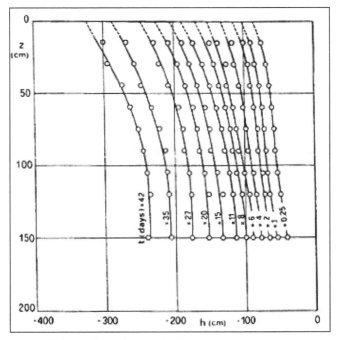

Fig. 16. Field measured h(z,t) values during drainage

The corresponding $\theta(z)$ profiles show marked discontinuities due to soil stratification
(Figure 17). In the layer closest to the surface (0<z<90 cm), moderate variations of water
content occur with increasing depth. Such variations then decrease as the drainage process
evolves.

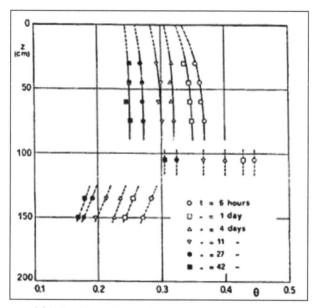

Fig. 17. Field measured θ(z,t) values during drainage

Profiles θ(z,t) for (0<z<90 cm) were used to obtain β and K_0 using each of the four simplified methods presented in equations (22), (24), (26), (29) and (30).

Table 2 reports both the parameters obtained by linear regression (equation (18)) among the average θ* water content of the profile and the water content 0 measured at depth z=75 cm, and the results of statistical fitting of equation (27) to the observations of average water content in the profile for different time values.

$\theta^*=a\theta+b$	a=0.9476	b=0.0132	R^2=0.9995
$\theta^*=At^B$	A=0.3539	B=-0.0830	R^2=0.9758

Table 2. Parameters of equations 18 (first line) and 27 (second line) for z=75 cm. R^2 is the coefficient of determination

Figure 18-19 indicate the high agreement between observed data and the estimated regression lines. Knowledge of parameters a, A and B, at the preselected depth of z=75 cm, then allowed us to determine with the various methods illustrated the analytical relations which supply K as a function of θ.

Profiles h(z) and θ(z) were then elaborated to obtain the conductivity K in function of θ by using the *instantaneous profile method*. Integration of the θ(z) profiles measured for assigned values of time t, together with evaluation, by means of experimental profiles h(z) relative to the same times, of hydraulic head gradients, allowed us to use Darcy's Law to obtain hydraulic conductivity values K, during evolution of the drainage process. The results of the elaborations performed for depth z=75 cm are shown in Figure 20 which also reports the diagrams of equation (5), obtained by assigning, at each step β and K_0 values calculated respectively by equations (22), (24), (26), (29) and (30) and reported in Table 3.

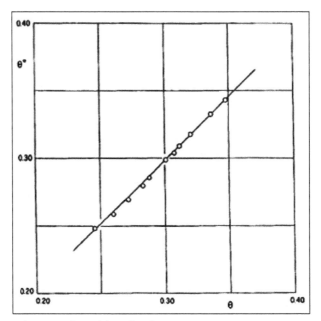

Fig. 18. Correlation between θ^* and θ at depth z=75 cm

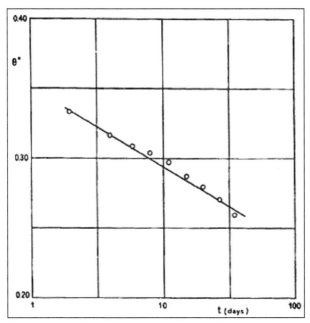

Fig. 19. Calculated curve (solid line) obtained by regression of equation 27 on measured
(symbols) of water content θ^* at depth z=75 cm

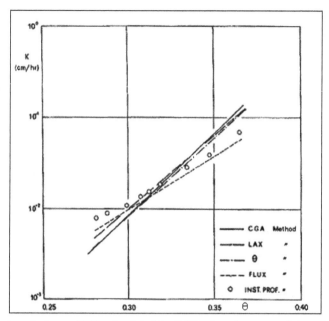

Fig. 20. Comparison between K(θ) calculated (solid line) and measured (symbols) at z=75 cm

Method	K_0 (cm/hr)	β	r
θ	1.12	37.26	0.981 [a]
Flux	1.06	27.21	1.986 [b]
Lax-θ	1.12	39.35	0.981 [a]
CGA	1.14	41.16	0.998 [c]

[a] Regression coefficient for $(\theta_0-\theta)$ versus ln$t$
[b] Regression coefficient for $\ln[aL(\delta\theta^*/\delta t)]$ versus $(\theta_0-\theta)$
[c] Regression coefficient for $\ln\theta^*$ versus ln$t$

Table 3. Values of parameter K_0 and β of the hydraulic conductivity function estimated by the simplified methods

Overall, the results obtained allow us to make the following comments on the soil in question:

- There is a satisfactory linear relationship between the evolution in time of moisture at the pre-selected depth and mean moisture θ^* in the profile. This confirms other experimental observations reported by Libardi et al. (1980) and by Nielsen et al. (1973) on Panoche soil.
- The unit gradient assumption, together with a postulated exponential link between K and θ (equation (19)), also consistently involves a linear relation between θ^* and $\theta(z)$.

However, no theoretical consideration allows us to justify a priori the validity of equation (27). Only the statistical fitting of the data allows us to judge the latter assumption. In our case the moisture drainage process in the profile is adequately described by equation (27).

This result also confirms those presented by Chong et al. (1981) for two Hawaiian soils with different hydraulic properties.

- The hypothesis of the unit hydraulic gradient, is not always verified (Ahuja et al., 1988). In our case Figure 21 shows, despite uncertainty due to measurement errors, that the gradient, initially close to -1, decreases slowly during the measurement period.
- It is worth noting that the various models may nevertheless also be used for hydraulic gradients different from -1 provided they are constant in time (Vauclin and Vachaud, 1987). The use of tensiometers allows us to evaluate the gradients and their possible fluctuations in time.

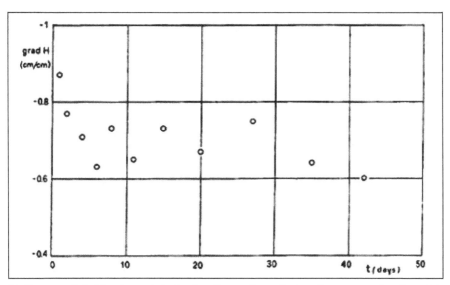

Fig. 21. Estimated field values of the hydraulic gradient during drainage at depth z= 75 cm

3.4 Disk permeameter method

Today, numerous mathematical models based on the solution of Richards' equation allow the numerical simulation of water and pollutant transfer in the vadose zone and thus may be essential instruments for assessing environmental pollution. However, the solutions offered by such models may contain considerable errors in the case of structured soils with heterogeneous pore systems which cannot be adequately described by the generally used unimodal retention and hydraulic functions. In such cases, for characterizing the flow regime, it is essential to separate the flow through macropores from the one through the soil matrix (Bouma, 1982). Recognition of this dichotomy is important because the properties of the macroporous system tend to dominate the infiltration process at and near saturation, while drainage, redistribution and root water uptake depend on those hydraulic properties which reflect the nature of the matrix.

Given the complexity of the problem, specific macropore models have recently been set up (Jarvis et al., 1991; Chen and Wagenet, 1991; Gerke and van Genuchten, 1993) which may more simply be considered as dual-porosity models. Implementation of such models requires new technologies and, because of the ephemeral nature of macropores, a new class of experiments to be conducted, if possible, in the field, so as to obtain input data for quantifying soil hydraulic properties, macropore distribution and spatial variation of macropores.

Various methods to obtain macropore parameters have been set up, such as tracer-breakthrough curves, computerized tomography, dye-staining and sectioning (Bouma and Dekker, 1978; Warner et al., 1989). However, in most cases such methods require undisturbed soil samples and they cannot be easily transferred to the open field. More simple methods, as recently suggested, are based on measuring unconfined infiltration rates with a disc permeameter (Perroux and White, 1988; Watson and Luxmoore, 1986; Ankeny et al., 1991) or a surface crust to restrict flow rates into the soil (Booltink et al., 1991). By offering slight hydraulic resistance to water movement, discs and crusts allow a water supply potential h to be applied at the soil surface, with a negligible head loss. By appropriately choosing the resistance value and ensuring that h is only -0.01 to -0.02 m, an acceptable approximation of soil hydraulic conductivity is obtained, which excludes the macropore system. It is thus possible, by the resolution of the 3-dimensional moisture flow field, to obtain soil hydraulic properties for water content values near saturation.

Finally, it is worth mentioning particular calculation methods proposed in the literature and set up for estimating hydraulic conductivity from disk permeameter data. Among others, the most commonly used are those proposed by White and Sully (1987) and by Ankeny et al. (1991). Such methods vary in complexity and simplifying assumptions, as well as having different advantages and limitations.

3.4.1 Disk permeameter

The apparatuses used to impose Dirichletís boundary condition $h_0 \leq 0$ at the soil surface, in which ho is the water supply potential, are called tension infiltrometers or disk permeameters in the literature.

Following the design of Perroux and White (1988), a permeameter (Figure 22) was set up by the Hydrology laboratory of the Dept. DITEC, University of Basilicata, Italy. The permeameter in question consists of a bubble tower which may be considered a Mariotte double regulator connected to a perspex disc 200 mm in diameter covered by a porous nylon membrane with air entry value of approximately 0.25 m. The first regulator allows us to apply via the membrane a constant water supply potential which can be controlled by adjusting the water level inside. The second regulator, which functions as a reservoir by means of a graduated scale, allows infiltration water volumes to be evaluated in time.

The experimental observations which may be carried out with such a device necessitate, between the base of the disc and the soil surface, that there is adequate hydraulic contact by means of a thin sand stratum previously wetted up to a volumetric water content close to 0.01.

A Review of Approaches for Measuring Soil Hydraulic Properties and Assessing the Impacts
of Spatial Dependence on the Results

125

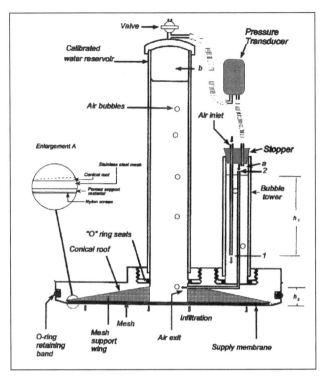

Fig. 22. Schematic illustration of the disk permeameter (Source:Evett et al., 1999)

3.4.2 Theory

The theoretical foundations for using a disk permeameter to infer soil hydraulic properties have been discussed in detail elsewhere (Perroux and White, 1988; Smettem and Clothier, 1988). Only the salient features are recalled below:

Sorptivity

The water flow emanating from a disk source according to Dirichlet boundary condition:

$h=h_0 \leq 0$; $z=0$; $t>0$

where h_0 is the water supply potential, z is the depth and t is the time, is initially controlled by soil capillarity (Philip, 1969):

$$\lim_{t \to 0} \left[\frac{Q(t)}{\pi r^2} \right] = \frac{1}{2} S_0 t^{-\frac{1}{2}}$$
(31)

in which Q is the flow rate ($L^3 T^{-1}$) from the disc source, t the time, r the radius of the source (L), $S_0 = S(h^0, h_n)$ the sorptivity ($LT^{-1/2}$,)and h_0 and h_n, respectively, the supply and initial water potential (L).

Integrating equation 31 with regard to t, we obtain:

$$I = S_0 t^{\frac{1}{2}} \tag{32}$$

in which I is the cumulated infiltration (L).

For lower time values starting from infiltrated water volumes, So may be simply deduced from the slope of I with regard to $\sqrt{t}$.

The geometric time scale of Philip (1969), t_{geom} may be used to evaluate when the geometric dominance of disc source should have been established.

This time is considered to be:

$$t_{geom} = \left[\frac{r\Delta\theta}{S}\right] \tag{33}$$

in which $\Delta\theta=\theta_0-\theta_n$ and θ_n are, respectively, water content values corresponding to the supply and initial water potential.

Steady state flow

The water flow rate will reach a stationary value henceforth termed Q_∞ for greater time values (Philip 1966).

In the case of multi-dimensional flow processes, Philip (1986) showed that a characteristic time scale t^* for the steady state flow rate is that for which the flow from the source is 1.05 times the steady flow rate. His calculations show that when the characteristic size of source r equals or exceeds the macroscopic capillary length scale, λ_c, an "innate soil length scale" as defined by Raats (1976), then it may be held that $t^*=t_{grav}$, with t_{grav} given by :

$$t_{grav} = \left(\frac{S}{k_0}\right)^2 \tag{34}$$

The above findings were recently confirmed by Warrick's studies (1992). Physically, t_{grav} represents the time in which the effects of gravity equal the effects of capillarity (Philip, 1969).

(Gardner, 1958) introduced the concept of "alpha" soils, that is soils whose hydraulic conductivity takes the exponential form $K = K_0 \exp(\alpha h)$, with α is a constant equivalent to λ_c^{-1} . In the case of "alpha" soils, Wooding (1988) showed that the steady flow from the source, for an assigned value of water supply potential, may be approximated with sufficient accuracy by the following expression:

$$\frac{Q_\infty}{\pi r^2} = \Delta K \left[1 + \frac{4\lambda_c}{\pi r}\right] \tag{35}$$

in which $\Delta K=K_0-K_n$.

Subsequently, White and Sully (1987) demonstrated that between the characteristic length scale k, sorptivity and hydraulic conductivity, there may be the following type of relation:

$$\lambda_c = \frac{bS^2}{\Delta\theta\Delta K} \tag{36}$$

in which b $(1/2 \leq b < \pi/4)$ is a shape factor frequently set at approximately 0.55 for agricultural soils.

If, as happens in many open field situations, it may be assumed that: $\Delta K = K_0 = K$ and if we substitute equation (36) in (35), the following simplified expression is obtained:

$$\frac{Q_\infty}{\pi r^2} = K + \frac{2.2S^2}{\Delta\theta\pi r} \tag{37}$$

An alternative approach based only on Wooding's solution is possible when we know the flows Q_∞ corresponding to the different r values of the source (Scotter et al., 1988, Smetten and Clothier, 1989). This approach allows K and λ_c to be evaluated directly, thereby solving two type (35) equations.

The same principle may be applied for a single r of the disc source but with infiltration measurements at multiple potentials (Ankeny, 1991). In particular, in this case, if the flow is measured with a single disc source at two pre-established potential values h_1 and h_2, two type (35) equations are obtained, which may be solved simultaneously, thereby supplying:

$$
\begin{aligned}
K_1 &= \frac{Q_1}{\pi r^2 + 2\Delta hr\left(1 + \frac{Q_2}{Q_1}\right)\left(1 - \frac{Q_2}{Q_1}\right)} \\
K_2 &= \frac{Q_2 K_1}{Q_1} \\
\lambda_c &= \frac{\Delta h(K_1 + K_2)}{2(K_1 - K_2)}
\end{aligned}
\tag{38}
$$

in which Q_1, and Q_2 are, respectively, Q_∞ at h_1, and h_2.

3.4.3 A case study

Infiltration tests were conducted at the experimental farm of the University of Basilicata near Corleto (Potenza, Italy) on bare soil which had undergone minimum tillage during the winter. The dominant soil at the experimental site is sandy clay (sand 36.5%, silt 24.1%, clay 39.4%).

From the pedological point of view, the soil may be classified as. "vertic ustorthens" according to the USDA classification system. Other important properties comprise moderate permeability and the vertic traits of clayey land which cracking renders quite appreciable during the summer.

At the test location one (1x1 m²) plot was isolated and subdivided into 4 equal subplots of 0.5x0.5 m². To determine soil hydraulic conductivity with White and Sully's method (1987), three water supply potential were used (-0.02;-0.06;-0.10 m), each in a single subplot.

According to the Ankeny et al. method (1991), in the 4th subplot the disc permeameter was not moved for trials at the three potentials (-0.06;.-0.04;-0.02 m). With this method, steady state flows were measured in an ascending sequence of supply potentials: first with a -0.06 m water supply potential followed by -0.04 and -0.02 m.

In each subplot, three undisturbed soil samples, 4.8 cm in diameter and 2.8 cm in height, were taken from the soil surface before and after the infiltration tests in order to measure bulk density and volumetric surface water content θ corresponding to h_0 and h_1. The mean dry bulk density of the first 2.8 cm of soil measured on undisturbed soil samples was estimated at 1.198 g/cm³ and the standard deviation at 0.05 g/cm³.

Following disk permeameter measurements, the soil was allowed to drain for about three hours and then an undisturbed soil sample 15 cm in diameter and 20 cm high was dug out of the 4th subplot immediately below the porous disk, to measure the hydraulic conductivity in the laboratory with the crust method (Figure 23)(Booltink et al., 1991; Coppola et al., 2000).

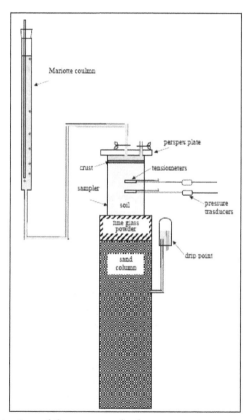

Fig. 23. Schematic illustration of the crust method apparatus

This method requires a pedestal of soil and a Stackman sand filter for draining the soil samples slowly saturated from the base. Conductivity values are determined during steady vertically downward flow under unit hydraulic gradient measured with small tensiometers. Once it has been ascertained that the water flux density in is equal to the water flux density out, the hydraulic conductivity will be equal to the imposed water flux. Moreover, it is worth noting that under the condition of unit hydraulic head gradient the matric potential at different depths is uniform, the water content is fairly uniform and, consequently, the accuracy in the estimate of K will only depend on measurement errors. Thus, the function K(h) of the soil in question was deduced with great accuracy, applying a series of steady water flux densities for an extended period of time by means of a porous disk connected to a bubble tower to control the water supply potential in a field of variation between 0 and -0.20 m and evaluating the hydraulic gradient from the tensiometer measurements at depths of 0.025 and 0.075 m.

White and Sully (1987) calculate sorptivity by means of equation (32). By contrast, Q_∞ is estimated for greater time values and the same authors use equations (36) and (37) to calculate the hydraulic conductivity. Thus, in using equation (36) $\Delta\theta$ is still required.

The main theoretical assumption underpinning the method is that cumulative infiltration I for lower t values varies linearly with the square root of time and, for greater values, linearly with t. Figures 24(a) and 24(b) show that for the soil in question the features of flow theory for disc permeameters are satisfied. Indeed, the above Figures clearly show that for the 3 different water supply potentials, data quality is sufficient and that data are always widely available to determine So by regression of I against t at the straight line portion of Figure 24(b) for t values from 0 to roughly 100 s. On the other hand, Q_∞ may be evaluated by regression of $I(t)$ against t for the long straight portion of Figure 24(a).

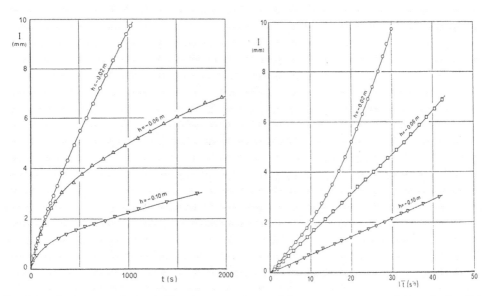

Fig. 24. a. In situ cumulative infiltration over time and b. over square root of time during 3D flow from disk permeameter for three water supply potentials

In the observed situation, it seems reasonable to expect a steady flow rate within less than an hour, which allows sufficient Q_∞ data to be obtained most rapidly. This last consideration is interesting in the case where it is intended to conduct inquiries to ascertain the level and pattern of spatial or temporal variability.

Amongst the methods based on Wooding's equation (35), the method proposed by Ankeny et al. (1991) is based only on measurements of steady-state flow rate Q_∞. However, determining when Q_∞ is reached may be very difficult (Warrick, 1992).

With regard to Figure 25, for the soil in question the mean time for reaching steady-state flow for the -0.06 m water potential was 100 s, followed by 400s to attain steady-state flow at -0.04 m water potential and finally 650 s for -0.02 water potential. In the above case, Q_∞ is obtained by the regression of $I(t)$ against t from Figure 25 and equations (38) are used to calculate conductivity K . Unlike the White and Sully method, in the Ankeny method measurements of θ_0 and θ_n are entirely avoided.

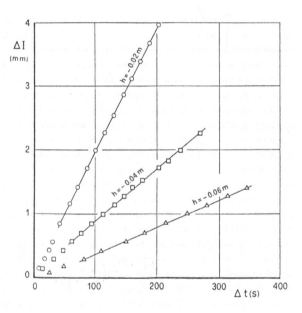

Fig. 25. Steady state flow intake for three water supply potentials

Conductivity values calculated by the various methods adopted are given in Figure 26, which also supplies, for the sake of comparison, laboratory core measurements of k based on the crust method. In particular, note that the K values calculated by the Ankeny method are in good agreement with the k values determined with the crust method. In any case, the bias ascertained for the White and Sully method to underestimate K values is within less than one order of magnitude.

The analysis of behaviour at the origin of function k(h) evidences a bimodal distribution of the porous system of the examined soil, with a water potential break-point at ≈-0.03 m. With

A Review of Approaches for Measuring Soil Hydraulic Properties and Assessing the Impacts
of Spatial Dependence on the Results

131

the increase in water potential from -0.02 m to 0, hydraulic conductivity increases by about an order of magnitude, nonetheless assuming numerical values greater than those measured in a previous measuring campaign on undisturbed soil samples taken in the same sites. Such behaviour suggests that the structural porosity of this clay soil may play a dominant role in determining the pattern of water flow in the field.

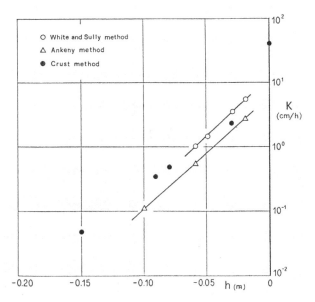

Fig. 26. Comparison of estimates of unsaturated hydraulic conductivity obtained using three different methods

Furthermore, to represent this bimodal pore system, a two-line regression model may be more responsive than the usual linear model, or than the model with several parameters proposed by van Genuchten (1980). Such findings were also reached by Messing and Jarvis (1993) in a similar pedological context.

Philip (1985), starting from macroscopic capillary length λ_c, infers a representative pore size λ_m (mm) by using the capillary theory:

$$\lambda_m = \frac{\tau}{\rho g \lambda_c} \cong \frac{7.4}{\lambda_c} \tag{39}$$

in which τ and ρ are, respectively, the surface tension and the water density, and g is the acceleration of gravity. The characteristic size λ_m defined by White and Sully (1987) to be a "physically plausible flow weighted pore size" may be considered a representative index of soil structure.

Starting from the measured values of S_0, K and $\Delta\theta$, estimates were made by means of equation (9) of λ_m, values which are supplied in table 4 together with the values of λ_c and t_{grav}.

h (m)	S (mm/s$^{1/2}$)	$\Delta\theta$	K (mm/s)	t_{grav} (h)	λ_c (mm)	λ_m (µm)
-0.020	0.193	0. 80	0.00697	0.17	13.1	572
-0.060	0.161	0.146	0.00154	3.04	64.4	118
-0.100	0.095	0.090	0.00035	21.20	122.6	61

Table 4. Values of physical and hydraulic properties of the examined soil

Examination of the table shows that the bimodal distribution is clearly evidenced by λ_m which shows a 9-fold change between -0.02 and -0.1 m water potential compared with a 5-fold change between -0.02 and -0.06 water potential and only a 2-fold change between -0.06 and -0.1 m water potential. At h>-0.06 m the influence of capillarity is very strong and t_{grav} is large. However, macropore flow increases as h approaches 0 and thus gravity flow is dominant at h<-0.06 m.

4. Spatial variability of soil physical properties

In light of the above, it appears evident that by improving measurement techniques and setting up mathematical models water flow in soil can be quantitatively determined. Moreover, since the transport in soils of solutes, pesticides and urban and industrial waste products, besides the reaction kinetics, is strictly linked to conductivity and hydraulic gradients, obviously only through knowledge of hydraulic characteristics is it possible to set up mathematical models that describe the environmental problems connected with such transport.

The results obtained from model applications depend on the quality of input data which, for field determinations, are greatly affected by the spatial variability of soils (Beckett and Webster, 1971). Soils vary considerably from place to place and only with accurate measurements can such variability be described. With pedological studies, soils can be classified and mapped, usually performed with a view to serving many applications and based on various factors, of which hydraulic properties play a minor role.

The various cartographic units have mean values with variations that are considered negligible, and depend on the classification scheme adopted and the scale of representation. Hence within the various cartographic units there are variations in hydraulic properties that cannot always be ignored. Applications of models to field conditions may lead to unacceptable errors when we consider soil as a homogeneous medium and, even more, when variations along the profile are considered, identifying layers in various horizons, each of which has mean values of functions $K(\theta)$ and $\theta(h)$. Thus soil has to be considered as a medium in which the properties are tied to spatial coordinates and, in the event of anisotropy, to direction.

Bearing in mind that soil extends over a large area in relation to its thickness and in the case of isotropy, hydraulic properties should always be considered, for each horizon, a function of two coordinates. A deterministic definition of spatial heterogeneity requires a large number of measurements and hence, owing to the burdensome, highly costly nature resulting from the complexity of testing techniques, cannot be pursued except for problems that affect limited zones. Moreover, flow processes should be viewed as three-dimensional, and their modelling would require high calculation costs with the use of computers with large memory capacities and high speeds, which are not always readily available.

Hence spatial variations in soil hydraulic properties are to be considered irregular, determined by innumerable parameters according to complex imperfectly known laws, which is why they appear essentially random and hence can be described only by means of statistical procedures. Working on an area of 150 ha, homogeneous in pedological terms, Nielsen et al. (1973) were the first to determine, at 20 randomly chosen sites and at six depths, the hydraulic conductivity curves $K(\theta)$ and those of tension $\theta(h)$, together with other physical parameters of the soil. The data show considerable variability from zone to zone and allowed, for each quantity measured, the corresponding frequency distribution to be identified.

The avenue pursued by Nielsen was then followed by others (Coelho, 1974; Carvallo et al., 1976; Kutilek and Nielsen, 1994; Coppola et al., 2009), who carried out surveys to characterise the heterogeneity of areas of different size. The data show marked variability also with reference to fairly small areas, especially as regards hydraulic conductivity and diffusivity which show variation coefficients in excess of 100%.

4.1 Simultaneous scaling analysis of soil water retention and hydraulic conductivity curves

The description of the spatial variability of hydraulic parameters may also be simplified by hypothesizing that soil microgeometry in two different sites is similar. As shown by Miller and Miller (1955a, b), values of water potential and conductivity measured in different zones may be related to corresponding averages through spatial distribution of the local similarity ratio a.

The validity of this hypothesis has been confirmed only through laboratory tests on sand filters (Klute and Wilkinson, 1958; Elrick et al., 1959) and, even if it finds no actual correspondence in soils, it has been used by several authors (Warrick et al., 1977b; Simmons et al., 1979; Rao et al., 1983) to relate potential and conductivity to the degree of soil saturation s and not to water content θ. The same authors evaluated the distribution of the similarity ratio not with reference to the microgeometry of the porous medium, but by ensuring that there was the best possible fit between the tension and conductivity curves, measured at various points.

Thus the rigorous concept of geometric similarity introduced 30 years ago by Miller and Miller is superseded and the similarity ratio is determined using regression techniques. The purpose of such techniques, generally indicated in the literature as being *functional normalization* (Tillotson and Nielsen, 1984), is to reduce dispersion of test data, concentrating them on an average reference curve that describes the relationship in question. Scale factors α identified with such procedures often differ in each hydraulic property on which they are evaluated (Warrick et al., 1977b; Russo and Bresler, 1980; Ahuja et al., 1984).

Application of the similarity concept to the water budget of a small river basin or an irrigation area has recently proved promising (Peck et al., 1977; Sharma and Luxmoore,1979; Warrick e Amoozegar-Fard, 1979; Bresler et al., 1979). Using a simulation model, the latter authors examined the effects of spatial variability of soil hydraulic properties, expressed in terms of each stochastic variable α, upon the components of the water budget. The various components of the budget, simulated by attributing log-normal frequency distributions to a, are in excellent agreement with the experimentally measured values. The literature also indicates that field-scale tests to apply the similarity concept are still too few and in any case restricted to soils with reasonably similar morphology.

4.1.1 Theory

In order to reduce dispersion of the experimental data and define a single average curve both for function h(s) and for k(s), the above-cited similarity theory expounded by Miller and Miller was applied. *The latter showed that, assuming constant* surface tension and the kinematic viscosity coefficient of water, for two porous media with geometric similarity the following formula holds:

$$W_r = \alpha_{w,r}^p \overline{W} \tag{40}$$

which links a generic hydraulic property W_r, determined in a site r, to the value $\overline{W}$ of the reference site. Factor $\alpha_{w,r}$ is the relationship between the lengths of the inner geometry of the medium at site r and those of the reference medium, while exponent p assumes the value -1, when the hydraulic property in question is the water potential h, and 2 when reference is made to hydraulic conductivity K. Under the similarity hypothesis the $\alpha_{w,r}$ coefficients are the same for any hydraulic property and the two sites have the same value of porosity. The values of coefficients a were determined for the area in question, expressing for each site r the relations linking the water potential h and hydraulic conductivity k at saturation s by means of assigned analytical functions.

It was postulated that one of the parameters appearing in the analytical relations, namely ar, depends on the measurement site, while the remaining ones, numbering m, assume the same values throughout the zone in question. Hence the following expression was used:

$$W_r = a_r f_w \left(s; b_1, \dots, b_m \right) \left(r = 1, \dots, n \right) \tag{41}$$

where n indicates the number of sites taken into consideration.

Having assigned the analytical expression for f, the parameters in question were determined by minimising the sum M of the square deviations between the values of the hydraulic property W, determined by function f_w, and the corresponding measured values $W_{r,j}$.

Hence to determine M the following expression was used:

$$M = \sum_{r=1}^{n} \sum_{j=1}^{n_r} \left[a_r f_w \left(s_r, j; b_1, \dots, b_m \right) - w_{r,j} \right]^2 \tag{42}$$

in which n_r is the number of points of curve W(s) relative to site r. Since the function f_w is generally non-linear, the minimum of M was sought with an iterative method similar to that proposed by Simmons et al. (1979) and a calculation program was specially written which, enabled rapid determinations starting from an initial estimate of parameters quite far from the solution.

Finally, the coefficients $\alpha_{w,r}$ defined by equation (40) were obtained by imposing the normalization condition:

$$\frac{1}{n}\sum_{1}^{n}\alpha_{w,r} = 1 \tag{43}$$

4.1.2 A case study

The present section aims to ascertain the validity of the method in context. In the following, we will provide some results of the case study illustrated in the section 3.1.1. For all the samples analyzed in Figure 27, in the function of degree of saturation, s, the values of potential $\bar{h} = \alpha_{h,r}h$, are presented, while in Figure 28 the conductivity values $\bar{k} = \alpha_{k,r}^{-2}k$ are given, obtained by processing the potential and conductivity data respectively.

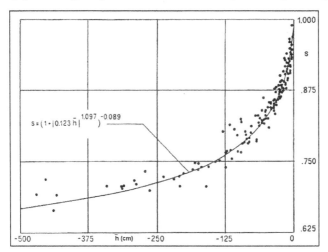

Fig. 27. Scaled soil water retention data for the composite

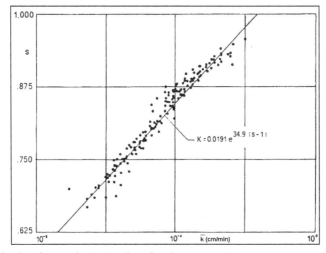

Fig. 28. Scaled hydraulic conductivity data for the composite

The graphs show how the method we adopted markedly reduces the dispersion of experimental points and allows the evaluation of the variability of the hydraulic properties of soil through the spatial distribution of the sole parameters $a_{h,r}$ and $a_{K,r}$.

The statistical processing thus demonstrated how the log-normal rule is that which best approximates to frequency distribution, not only to the parameters $a_{h,r}$ but also to those $a_{K,r}$. The principal statistical moments of $a_{h,r}$ and $a_{K,r}$ are given for the applications in table 5.

	$A_{h,r}$	$a_{K,r}$
mean	0.978	1.098
standard error	0.924	1.968
CV	0.945	1.793

Table 5. Values of mean, standard deviation and coefficient of variation of the scale factors determined respectively from h(s) and K(s) data

A study of the table reveals that the coefficients of variation obtained by processing the conductivity data are higher than those relative to the processing of the potential data, which is probably due to the fact that conductivity values are more sensitive to degrees of saturation. It must, finally, be underlined that the high degree of correlation between parameters $a_{h,r}$ and $a_{K,r}$ supports the macroscopic similarity hypothesis of Miller and Miller (1955).

4.2 Geostatistical analysis

A survey of the recent literature shows that various techniques are adopted to determine the spatial variability structure of soil hydraulic prperties. The possible periodicity of data may be highlighted through the spectral density function determined by adopting smoothing techniques to remove any errors to remove any errors (Webster, 1977). Under the assumption that soil parameters may be represented by random stationary functions, with finite mean and variance, the autocorrelation function can be defined as dependent only on the distance between observation pairs. The diagram of the autocorrelation function then allows us to determine up to what distance the observations are correlated with one another (Vauclin et al., 1982; Webster and Cuanalo de la C., 1975).

Recently, geostatistical techniques developed originally by Matheron for mining estimates (David, 1977; Journel and Huijbregts, 1978) have been profitably used to study the spatial variability of soil physical and hydraulic properties.

4.2.1 Theory

Like natural phenomena which, over space or time, show a structure and in geostatistics are termed regional phenomena (Matheron, 1971), by the same token hydraulic parameters may be considered regionalized variables. Indeed, with the use of stochastic procedures developed in the context of geostatistics, soil hydraulic parameters, actually functions with values distributed in space, come to be represented as stochastic variables Z, with discrete values assigned according to probability distributions laws.

At each point the properties of Z thus remain defined through the mean value, variance and covariance, with means referring to the ensemble of realizations at the point considered.

For practical purposes, it is nonetheless necessary to make some simplifying assumptions, since the probability distribution functions at each point in space are not known, nor can they be deduced in the availability of a single realization of the parameter in question.

Reference is therefore made to the intrinsic hypothesis and the stochastic variable is assumed to remain defined through a single function at all points. In relation to increases in Z, $Z(x+\overline{h}) - Z(x)$, for the assumed stationary conditions, there is a zero mean and variance independent of x. Moreover, in the availability of observations concerning distinct points being considered a single realization in space, with reference to the ergodic hypothesis it is assumed that the mean values obtained on the ensemble realizations for the single points may refer to spatial mean. Under such hypotheses, the variogram function remains definite:

$$\gamma(h) = \frac{1}{2}E\left[\left\{Z(x+\overline{h}) - Z(x)\right\}^2\right] \tag{44}$$

and again for the covariance, C(h), and variogram, $\gamma(\overline{h})$, functions only of vector $\overline{h}$, usually known as lag, the following condition is satisfied:

$$\gamma(\overline{h}) = C(0) - C(\overline{h}) \tag{45}$$

in which C(0) is none other than the variance of the process Z(x) assumed stationary and ergodic.

Upon verification of such hypotheses, the variogram function describes the structure of a regional variable of known value in a discrete number of points mostly distributed irregularly. Complete interpretation of the variogram function may be found in Journel and Huijbregts (1978). However, in short, with reference to a typical semivariogram, it may be stated that in the cases in which γ(h) is constant upon the variation in h, the values of the parameter in question may be considered spatially uncorrelated.

When, as h increases, γ(h) grows towards a constant value $C_1+C_0=C$, the observations are correlated amongst themselves as far as a distance h_0 at which the variogram may be held to have reached C. The maximum value C of the variogram is indicated as the sill and typically supplies an estimate of variance, while the distance h_0 at which γ(h) first reaches the sill value is indicated as a range. In theory, for h=0 we should have γ(0)=0, but the extrapolated test variograms that make h tend to zero are approximated to a value C_0 different from zero, called nugget effect which is indicative of the variability of the parameter at a smaller scale than the minimum sampling distance and the random errors tied to the measuring techniques used.

Once the spatial dependence of the observations has been identified, geostatistical techniques allow the burden of measurements to be contained by determining the minimum number of measurements, the optimal distance and their location. Moreover, with knowledge of the variogram function it is possible to apply optimal interpolation techniques (kriging) which allow fairly detailed spatial distribution maps of the various parameters to be drawn with a smaller number of observations (Burgess and Webster, 1980).

4.2.2 A case study

To show the presence of a spatial structure of the soil hydraulic properties, below we will analyze data coming from the experiment illustrated in section 3.2.2. To express the law of retention θ(h) we adopted the analytical relation proposed by van Genuchten (1980):

$$\theta(h) = \theta_r + \frac{\theta_s - \theta_r}{\left[1 + \left(\alpha|h|\right)^n\right]^m} \tag{46}$$

in which α and n are essentially empirical parameters, θ_s is the value that θ assumes at saturation, and θ_r represents residual water content, that is, what is found for h tending to minus infinity.

For the conductivity curve the exponential relation (Libardi at al, 1980) was used:

$$K(\theta) = K_0 \exp\left[\beta(\theta - \theta_s)\right] \tag{47}$$

in which β is an empirical parameter and k_0 represents the value of hydraulic conductivity for $\theta = \theta_s$.

Of the parameters which feature in relations (46) and (47), θ_s was determined with sufficient accuracy by means of measurements made at the end of the infiltration process, while θ_r was set at zero. Soil hydraulic properties are thus defined by the values of only four parameters: α, n, K_0 and β.

Table 6 gives a synopsis of results of all the determinations using exploratory statistics: mean, standard deviation and coefficient of variation. The physical properties of the soil in

		Statistical indices		
		μ	σ	CV (%)
Physical properties:	Sand (%)	80.0	3.84	4.80
	Silt (%)	12.0	0.68	5.67
	Clay (%)	8.0	0.26	2.06
	ρ_b (g/cm^3)	1.16	0.05	4.31
Hydraulic properties:	θ_s (-)	0.526	0.047	9.1
	θ_{-100} (-)	0.419	0.041	9.79
	θ_{-200} (-)	0.346	0.043	12.4
	θ_{-300} (-)	0.301	0.046	15.3
	Ks (cm/h)	8.31	3.75	45.13
	$K_{0.4}$ (cm/h)	0.051	0.031	60.78
	$K_{0.3}$ (cm/h)	0.0025	0.0018	72.00
	$K_{0.2}$ (cm/h)	0.00013	0.0001	76.92
Parametric curve coefficients	α (cm^{-1})	0.01	0.0026	26.00
	n (-)	1.46	0.139	9.52
	β (-)	29.9	5.56	18.60
	K_0 (cm/h)	0.50	0.48	96.0

Table 6. Exploratory statistics for some physical and hydraulic properties and parametric curve coefficients

question present a modest dispersion against the mean, with coefficients of variation on average equal to 4.21%. Also the coefficients of variation of water contents θ, for h=0, -100, -200 and -300 respectively, are somewhat low and on average equal to 11.6%. Moreover, as with the findings reported by Nielsen et al. (1973), the hydraulic conductivity values experience considerable variability with a coefficient of variation exceeding 60%. Finally, on examining the data for the four parameters represented in equations (46) and (47), the distributions of values α and β present dispersion around the mean greater than that of n, while the coefficient of variation of K_0 is 96%, higher than that observed for the conductivity K_s measured experimentally in the field.

Comments on the data in table 6 implicitly assume that the observations of the physical and hydraulic properties of the soil in question are independent of the point at which they were made. In reality, each property is never distributed in a disordered way: a certain spatial structure is always present.

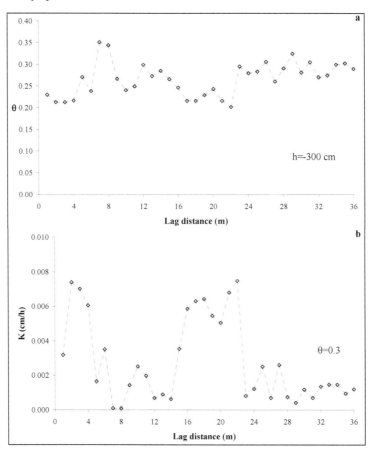

Fig. 29. a. Series of θ_i values, i=1, 2,…36 obtained from water content measurement for h=-300 cm; b. Series of K_i values, i=1, 2,…36 obtained from estimation of hydraulic conductivity for θ=0.300

To show the presence of a spatial structure for hydraulic properties of the soil in question, below in Figure 29a, b we provide spatial distributions of water content and hydraulic conductivity for the individual soil profiles along the transect at a respective depth of z=30 cm and z=45 cm.

It is inferred that θand K values show a discrete variability along the transects definitely tied to the local characteristics of the porous medium with deviation from the mean heavily dependent on the measuring point. The structure of spatial variability was then quantitatively evaluated using geostatistical techniques. The values of the experimental semivariograms γ for the data in Figure 29a, b are reported respectively in Figures 30a and 30b as a function of Lag, in other words the number of intervals between observations. The diagrams clearly show the presence of a spatial structure with values related between them up to a distance of about 7-9 m. Moreover, on fitting the diagrams for h=0, one notes that $\gamma(0) \neq 0$. This pattern that is often encountered in applications denotes the presence of noise components in the random function that may, in this case, be explained by random fluctuations, measurements errors and scale effects.

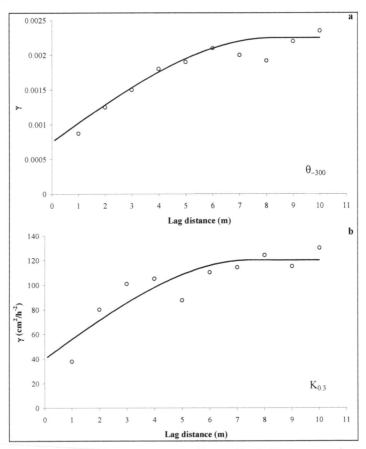

Fig. 30. a. Variogram relative to water content in Figure 29a; b. Variogram relative to conductivity in Figure 29b

The spherical model, chosen among the so-called authorised models, was then used to describe the variograms in Figure 30a, b. The iterative estimates, obtained with the least squares method from the model parameters, supplied the results of table 7. The last column of the table shows that 67% and 33% of the variability of θ and K is spatially structured.

Parameter	Field (m)	C_0	C_1	$100 * C_1 / (C_0 + C_1)$
θ_{-300}	8.25	0.00075	0.0015	66.7
$K_{0.3}$	7.50	40	80	33.3

Table 7. Parameter estimation of the spherical model used for variograms of Figure 30a,b

4.3 State-space analysis

Another group of techniques, also used to study the structure of variability, is based on a modified time series theory (Box and Jenkins 1970). By using such techniques, the structure may be described in terms of autocorrelation functions and SARMA (Spatial Autoregressive-Moving Average) models with a view to estimating the stochastic properties of the data. Some of these applications in soil physics and hydrology include the studies by Morkoc et al., (1985); Anderson and Cassel, (1986); Wendroth et al., (1992); Cassel et al., (2000); Heuvelink and Webster (2001), Wendroth et all. (2006).

4.3.1 State-space model formulation

Clearly, there is a strict analogy between space and time, at least in the case of one-dimensional space. Hence, under the hypothesis of isotropy, analytical methods are to a broad extent equivalent. Typically, time series analysis allows us to analyse spatial structure in terms of autocorrelation functions and generalisation of state-space models. For this particular method of regression in the time and space domain, unlike the methods of kriging and cokriging (Vieira et al., 1983) the assumption of stationarity of observations is not required. The state-space method (Kalman, 1960) is particularly interesting when the phenomenon in question satisfies certain systems of differential equations. The method has been used in economics (Shumway and Stoffer; 2000) and has yielded good results in agronomic and soil science (Vieira et al., 1983; Morkoc et al., 1985; Wendroth et al., 1992; Comegna and Vitale, 1993; Wu et al., 1997; Cassel et al., 2000; Poulsen et al., 2003; Nielsen and Wendroth, 2003)

Let us use $Y(x)$, $x = x_o + 1$, ..., $x_o + n$, to indicate the values assumed by n observations made for a certain soil parameter Y along a given transect (below we shall use the simpler notation Y_t, $t = 1, 2, ..., n$). A state-space model consists, in the formulation most useful for our purposes (for details and generalisations see Anderson and Moore, 1979), of two equations:

$$\begin{cases} Y_t = F_t' Z_t + v_t \qquad Z_t = G_{1t} Z_{t-1} + G_{2t} Z_{t+1} + w_t \\ Z_t = G_t Z_{t-1} + w_t \Rightarrow \qquad\qquad\qquad\qquad t = 1, 2, ... n \end{cases} \qquad (48)$$

$$\text{isotropic} \qquad\qquad\qquad \text{anisotropic}$$

The first termed that of observations and the second that of transition, where F_t is a known vector (p, 1), Z_t is a vector (p,1) of the system state, G_t, G_{1t}, G_{2t} are a set matrices (p, p);

$v_t \sim N(0; \sigma^2_{v_t})$ independent of $w_t \sim Np(0; \Sigma_{w_t})$. The model (48) is wholly specified by the parameters $(F_t, G_{it}, \sigma^2_{v_t}, \Sigma_{w_t})$ and includes, as particular cases, other statistical models such as regression, ARIMA and SARMA models.

Having set the initial values, we may obtain optimal forecasts and estimates of the non-observable components by using the *Kalman filter*. At the same time, from many observations made of soil physical and hydraulic properties, the latter may plausibly have been generated by stationary *isotropic processes* with parameters independent of the individual measuring points:

$$E(Y_t) = \mu; \quad var(Y_t) = \sigma^2 \quad cov(Y_t, Y_{t \pm h}) = c(h)$$

Hence we may consider the case in which the equations in (48) are reduced to simple ARMA and SARMA models. The importance of being able to make the double representation (state-space and SARMA or ARMA) lies in the fact that ARMA and SARMA models are easy to identify and estimate, while state-space models allow a more straightforward, immediate interpretation of the phenomena to which they are applied. Indeed, from (48) it follows that Y_t may be interpreted as the result of signal $F_t' Z_t$ which is overlaid by a random error v_t. Evolution of many physical phenomena can be well represented with a logical scheme like that reported in Figure 31.

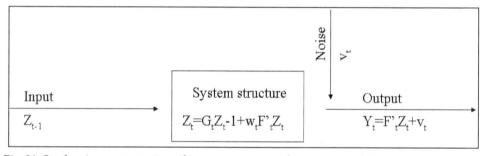

Fig. 31. Stochastic representation of input-output transformation model

The system structure is usually very straightforward and can be approximated, in the isotropic case, by an AR(1), given by:

$$Z_t = \phi Z_{t-1} + w_t$$

or, in the anisotropic case, a SAR(1) given by:

$$Z_t = c_0 + \phi_1 Z_{t-1} + \phi_2 Z_{t+1} + w_t$$

Note that, if it is $\phi_1 = \phi_2$ then the SAR(1) model may be replaced by the simpler AR(1) model.

Moreover, if we assume $p=1$, $F_t = 1$, $G_t = \phi$ then we obtain more simply:

$$\begin{cases} Y_t = Z_t + v_t \\ Z_t = \phi Z_{t-1} + w_t \end{cases} \Leftrightarrow \begin{cases} (-\phi B)Y_t = (-\alpha B)e_t \\ (1 - \phi B)Z_t = w_t \end{cases}, \quad t = 1, 2, ..., n \qquad (49)$$

A Review of Approaches for Measuring Soil Hydraulic Properties and Assessing the Impacts
of Spatial Dependence on the Results
143

where B is the backshift autoregressive operator, $\phi_i > a_i$ and e_t such that $e_t - a_1 e_{t-1} - a_2 e_{t-1} = v - \phi_1$ $v_{t-1} - \phi_2 v_{t+1} + w_t$. Thus both the equation of the observations and that of transition (i.e. the signal) are reduced to simple ARMA models and especially to an ARMA (1,1) for Y_t and an AR(1) for Z_t.

4.3.2 A univariate case study

In this section, we will still refer to the data set coming from the experiment shown in section 3.2.2, by analyzing individually the two parameters which characterize the soil water status in terms of θ and h measured at 0.3 m depth, along the N-S line of the plot so as to highlight their intrinsic structure linked to regional variability and, for 3 of the 12 measuring sampling times (the 3rd, 6th and 11th carried out 48, 168 and 768 hours respectively from the start of the drainage), the variations occurring in time (the parameters concerned are indicated by θ_i and h_i).

The data were first elaborated using classical statistical techniques, hypothesizing that the parameters vary in an essentially random manner. From this point of view, the main statistical indices (min. value, max. value, mean, standard deviation, skewness, kurtosis, coefficient of variation) of the above parameters are reported in table 8.

	Min	Max	Mean	SD	Skew	Kurt	CV
θ_3 (-)	0.307	0.383	0.341	0.019	0.217	-0.675	0.056
θ_6 (-)	0.257	0.330	0.287	0.018	4.462	-0.505	0.062
θ_{11} (-)	0.205	0.283	0.239	0.016	0.256	-0.353	0.068
θ_{-100} (-)	0.236	0.300	0.257	0.015	0.655	-0.233	0.057
h_3 (cm)	58.0	103.1	83.2	10.4	-0.411	-0.217	0.125
h_6 (cm)	113.1	180.9	147.3	16.4	0.141	-0.700	0.111
h_{11} (cm)	189.7	305.2	244.9	27.8	0.108	-0.670	0.113

Table 8. Main statistical indices of the analysed series

From table 8 we may deduce, for all the measuring times considered, an increase in the standard deviation (SD) with its mean for parameter h, whereas the SD of θ is practically constant. We also note that the coefficient of variation (CV) of h is almost twice that of θ. Concluding, the two processes describing h and θ are, in time, both non-stationary on the mean, while h is also non-stationary in variance. Figure 32 illustrates the above points: it reports the 50 observations of θ and h for 10 of the 12 sampling times (from the 2nd to the 11th). This all agrees with the theoretical results obtained by Yeh et al. (1985), which predicted such behaviour on the basis of the stochastic analysis of unsaturated flow through heterogeneous media.

In the context of stochastic analysis it is essential to verify, for the parameters considered, the existence of a correlation structure.

Variables θ and h, given that they are recorded at constant intervals along the transect, are ordered in space and their evolution in the prefixed direction can therefore be evaluated by means of typical statistical analyses of the time series and in particular by means of the ARMA model (see Box and Jenkins, 1970) only when the hypothesis of isotropy can be justified. A preliminary test was than carried out on θ_3 and h_3 series. The model which supplied the most acceptable results in terms of simplicity and interpretability, both because of the limited number of parameters and goodness of fit of the data, was SAR(1) model where the unknown parameters to be estimated are ϕ_1 and ϕ_2. The above criteria of choice was followed for all models subsequently used. It should be noted that if ϕ_1 and ϕ_2 are substantially equal than θ_3 and h_3 are to be considered isotropic, conversely anisotropy may be taken into account.

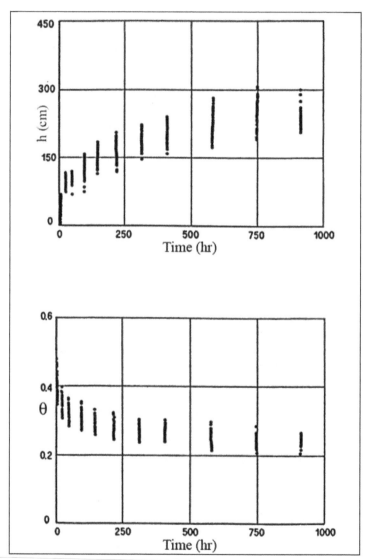

Fig. 32. Soil water potential h and volumetric water content θ as a function of time at the 0.3 m depth for all redistribution times during the drainage period

The iterative least-squares method estimate of the model parameters in question, provided the results reported in table 9 where standard deviations are in brackets. Having supposed that the phenomenon is isotropic and therefore invertible in space, the estimated ϕ_1 and ϕ_2 values are expected to be equal. In particular, in our case, we may observe that $\phi_1 = \phi_2$.

	θ_3	h_3
c_0	0.0858	29688.27
	(0.0418)	(12629)
ϕ_1	0.3950	0.331
	(0.1409)	(0.1341)
ϕ_2	0.3530	0.317
	(0.1488)	(0.1374)
σ_w	0.01435	8935.8
R^2	0.4568	0.2907

Table 9. Parameter estimates and comparison of SAR(1) model for the series examined and goodness of fit index R^2; in parenthesis the standard deviation of the estimates

If the estimates are analyzed in greater detail, in the Figures 33a,b,c it may be noted that the parameters in question are statistically identical. Then the model reported in (48) was applied, only to three series of data obtained along the transect. The series concerns, in particular, values of soil water content θ and tension h obtained 48 hours from the beginning of the drainage. Furthermore the analysis will be extended to a section of the soil moisture retention curve $\theta(h)$ constructed for h=100 cm, subsequently indicated as θ_{-100}. More significantly, the essential characters assumed in the space from the distribution of the parameters in question may be deduced from the transects of Figure 33, which report the relative values in the 50 observation points.

To identify the ARMA models to be adapted to the above three series, we estimated the autocorrelation (ACF) and partial autocorrelation (PACF) function. As transpires from Figures 34a,b,c,d, the three series can be well represented by an ARMA (1,1) model.

Anyway analysis of ACF residuals (Figure 34d) shows clearly that no structure whatever is present in the series of noises, which is further confirmation of the good fit of the model used to represent the examined parameters.

Parameter estimates, obtained with the least squares method, of the ARMA(1,1) model adapted to the above series and the goodness index fit R^2 (mean square errors in brackets) are reported in table 10. Clearly, all three series show a strong inertia component which confirms the presence of the spatial structure ascribable to an AR(1), accompanied by marked fortuitousness as results from the low value of R^2.

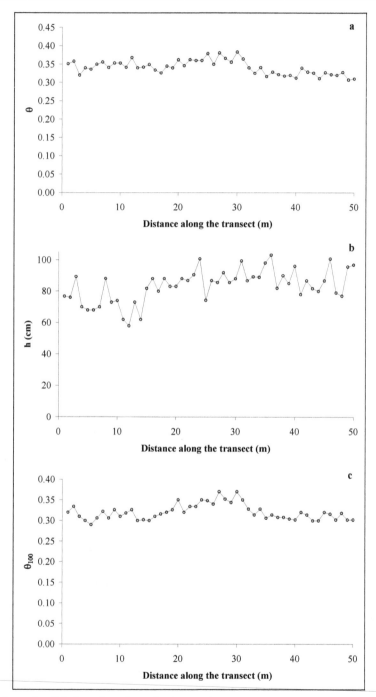

Fig. 33. Values of a. θ, b. h, c. θ_{100} along the transect

A Review of Approaches for Measuring Soil Hydraulic Properties and Assessing the Impacts
of Spatial Dependence on the Results

147

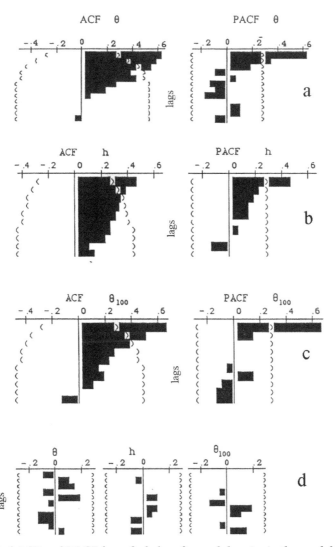

Fig. 34. Estimated ACF and PACF for a. θ, b. h, c. θ_{100} and d. noise in the model 1

	ϕ	α	$\hat{\sigma}_l$	R^2
θ_3	0.94 (0.075)	0.51 (0.12)	0.0136	0.501
h_3	0.91 (0.07)	0.63 (0.14)	8.729	0.300
θ_{100}	0.77 (0.10)	0.21 (0.10)	0.0110	0.392

Table 10. Parameter estimates of the ARMA(1,1) model, for θ_3, h_3, θ_{-100} and goodness of fit R^2. In parenthesis the standard deviations of the estimated parameters.

The estimated model was then used to obtain optimal predictions along the transect. Figure 35a, b reports the observed data, a signal estimate and the relative noise for series θ_3 and h_3. The graphs for the other series were similar, with analogous signal in the general pattern, not reported here for the sake of brevity, confirming that the spatial structure of soil hydraulic parameters is a characteristic of the porous medium in question.

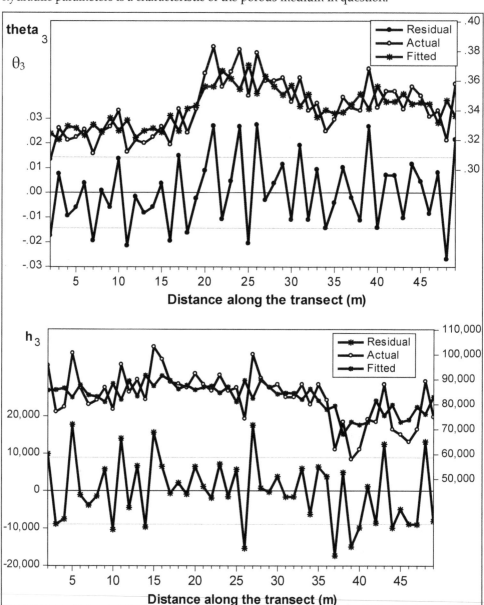

Fig. 35. Observed data, signal estimate and the relative noise for series a. θ_3 and b. h_3.

A Review of Approaches for Measuring Soil Hydraulic Properties and Assessing the Impacts
of Spatial Dependence on the Results
149

5. Conclusions

A determination of the hydraulic properties of soil is seen to be indispensable if we wish to approach the study of water movement in quantitative terms, using mathematical models. The practical possibilities of extensive application of such models implies the development of measuring techniques which render the determination of the laws θ(h) and K(θ) less problematic.

These proposed methods allow us readily to obtain functions θ(h) and K(θ), by submitting the samples, in the laboratory and in the open field, to a process of evaporation and internal drainage, respectively, and requires only the measurement of the mean water content and of the potentials at different depths along the soil sample or profile.

It was verified that the retention and conductivity curves thus determined and described, using analytic equations, can be used in simulation models with an acceptable degree of accuracy. Moreover, it can be observed that the duration of the tests was not excessive: this makes acceptable the labour intensiveness of characterization imposed on statistical evaluations of the variability of individual hydraulic parameters.

Therefore, from this proven association of accuracy and speed in the experimental work of characterization, we are convinced that the adopted methodologies can be considered adequate to the demands of modern study at an operative level of problems connected with water movement and of solutes in unsaturated media.

6. Appendix A

6.1 Hydraulic conductivity function

Hydraulic conductivity (K) is one of the most complex and important of the properties of vadose zone in hydro-physics and of aquifers in hydrogeology as the values found in nature:

- range of many orders of magnitude (the distribution is often considered to be lognormal);
- vary a large amount through space (sometimes considered to be randomly spatially distributed or stochastic in nature);
- are directional (in general K is a symmetric second-rank tensor; e.g.; vertical K values can be several orders of magnitude smaller than horizontal K values;
- are scale dependent;
- must be determined through field test, laboratory column flow test or inverse computer simulation;
- are very dependent in a non-linear way on the water content which makes solving the unsaturated flow equation difficult.

Some typical K(θ) curves for sandy and loamy soils are shown in Figure 1A. Note that at saturation the sandy soil has higher conductivity because of the larger water-filled pore spaces. These large pores drain at low suction, however, causing a considerable decrease in conductivity. The loamy soil, having smaller (on average) pores, retains a greater number of water-filled pores and thus has higher conductivity than the sandy soil at these potentials.

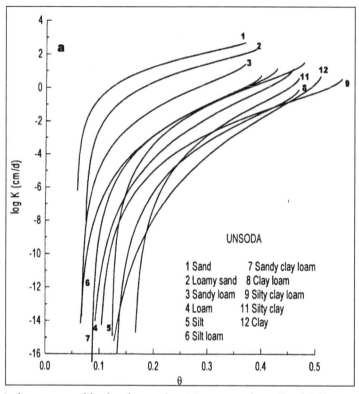

Fig. 1A. Typical unsaturated hydraulic conductivity curves for soils of different texture from UNSODA database. Source: Leij et al. (1996)

6.2 Parametric models for soil unsaturated hydraulic conductivity

In the past, many different functional relations have been proposed in the literature based on various combinations of the dependent variables θ, h and K, and a certain number of fitting parameters (Brooks and Corey, 1964; van Genuchten, 1980). Independent of the correct physical meaning of the fitting parameters, their values are submitted to constraints imposed by the use of the transfer equations such as the Fokker Planck and/or Richards equation. The most frequently used in the literature are the following water retention and hydraulic conductivity expressions:

Soil water characteristic functions

The Brooks and Corey (1964) equation:

$$\frac{\theta - \theta_r}{\theta_s - \theta_r} = \left(\frac{h_{bc}}{h}\right)^{\lambda} \qquad h \le h_{bc}$$

$$\theta = \theta_s \qquad\qquad h_{bc} \le h \le 0 \tag{1A}$$

and the van Genuchten (1980) soil water characteristic equation:

A Review of Approaches for Measuring Soil Hydraulic Properties and Assessing the Impacts
of Spatial Dependence on the Results

151

$$\frac{\theta - \theta_r}{\theta_s - \theta_r} = \left[1 + \left(\frac{h}{h_g}\right)^n\right]^m \tag{2A}$$

where the water pressure head (h) is usually taken as negative and expressed in cm of water; h_{bc} is the Brooks and Corey pressure scale parameter, h_g is the van Genuchten pressure scale parameter, and λ, m, and n are water retention shape parameters. The water retention shape parameters m and n are assumed to be linked by

$$m = 1 - \frac{k_m}{n} \quad n \geq k_m \tag{3A}$$

where k_m is an integer value initially introduced by van Genuchten (1980) to calculate closed-form analytical expressions for the hydraulic conductivity function when substituted in the predictive conductivity models of Burdine (1953) or Mualem (1976). For the Mualem theory, parameter k_m takes the value $k_m = 1$, and for the Burdine theory $k_m = 2$. For high pressure head values, the van Genuchten water retention equation behaves like the Brooks and Corey equation with $\lambda = mn$. However, it should be noted that this identity is only confirmed for soils with shape parameter values m <0.1 (for the case where the Burdine condition is used: $k_m = 2$);

Hydraulic conductivity functions

The Brooks and Corey (1964) equation:

$$\frac{K}{K_s} = \left[\frac{\theta - \theta_r}{\theta_s - \theta_r}\right]^\eta \tag{4A}$$

and the van Genuchten-Mualem (1980) hydraulic conductivity equation:

$$\frac{K}{K_s} = \left[\frac{\theta - \theta_r}{\theta_s - \theta_r}\right]^{1/2} \left[1 - \left\{1 - \left(\frac{\theta - \theta_r}{\theta_s - \theta_r}\right)^{1/m}\right\}^m\right]^2 \tag{5A}$$

where η is a conductivity shape parameter.

Through an extensive study, Fuentes et al. (1992) concluded that only the combination of the van Genuchten water retention equation (equation 2A), h(θ), based on the Burdine theory (m=1-2/n) together with the Brooks and Corey conductivity equation (Equation 5A) stays valid for all different types of soil encountered in practice without becoming inconsistent with the general water transfer theory. This is due to the rather limiting constraint which exists for shape parameter m when using the Mualem theory : 0.15 ≤m≤1 Even though the residual water content (θ_r) has a well-defined physical meaning, the parameter θ_r, which enters in equations (1A) to (5A) is somewhat of a misnomer because it usually behaves as a pure fitting parameter without any physical meaning. For practical purposes, it can easily be set equal to zero (Kool et al., 1987).

The scale parameter K_s is strongly related to soil structure. Among the different soil hydraulic characteristic parameters the saturated hydraulic conductivity is the parameter which is the most influenced by effects such as macropores, stones, fissures, cracks, and other irregularities formed for various biological and mechanical reasons. Hence, it is the parameter which is the most difficult to predict. The models proposed in the literature either give estimations of the capillary conductivity value or are based on site and soil specific databases. The results should therefore be considered with caution when applied to field studies.

Mishra and Parker (1990) used the Mualem model (1976) with the van Genuchten water retention function (equation 2A) to obtain a closed-form expression of the saturated hydraulic conductivity:

$$K_s = c_1 \frac{\left[\theta_s - \theta_r\right]^{2.5}}{h_g^2} \tag{6A}$$

where c_1 is a constant including the effects of fluid characteristics and the porous media geometric factor; it has a value of 108 cm³/s when K_s is expressed in cm/s; h_g is the van Genuchten (1980) pressure scale parameter. The authors derived a similar predictive equation by the use of the Brooks and Corey (1964) water retention equation (equation 1A):

$$K_s = c_1 \frac{\left[\theta_s - \theta_r\right]^{2.5}}{h_{bc}^2} \left[\frac{\lambda}{1+\lambda}\right]^2 \tag{7A}$$

where λ and h_{bc} are the Brooks and Corey (1964) shape and scale parameters, respectively. Ahuja et al. (1985) used the general Kozeny-Carman approach to determine the saturated hydraulic conductivity from the effective porosity ($\varepsilon - \theta_r$):

$$Ks = c_2 \left[\varepsilon - \theta_r\right]^{c_3} \tag{8A}$$

where c_2 is equal to 1058 cm/h when K_s is expressed in cm/h; and c_3 takes a value of 4 or 5.

Estimation of van Genuchten and Brooks and Corey parameters from experimental data requires sufficient data points to characterize the shape of the curves, and a program to perform non-linear regression. In particular, computer programs for estimation of specific parametric models are also available, e.g., the RETC code (van Genuchten at al, 1991).

7. References

Ahuja L.R., B.B. Barnes, D.K. Cassel, R.R. Bruce, D.L. Nofziger (1988), Effect of assumed unit gradient during drainage on the determination of unsaturated hydraulic conductivity and infiltration parameters, *Soil Sci.*, 145, 235-243.

Ahuja, L. R., Naney, J. W., and Williams, R. D. (1985), Estimating soil water characteristics from simpler properties or limited data. *Soil Sci. Soc. Am. J.* 49: 1100-1105.

Amoozegar-fard D., W.W. Warrick, Field measurements of saturated hydraulic conductivity (1986), In: a. Klute ed., methods of soil analysis, part 1: physical and mineralogical methods, monograph series 9. *Am. Soc. Agron.*, Medison; WI

Anderson S.H., D.K. Cassel (1986), Statistical and autoregressive analysis of soil properties of Portsmouth sandy loam, *Soil Sci. Soc. Am. J.*, 50, 1096-1104.

Ankeny M.D., M. Ahmed, T.C. Kaspar, R. Horton (1991), Simple field method for determining hydraulic conductivity, *Soil Sci. Soc. Am. J.*, 55, 467-470.

Basile A., Coppola A., De Mascellis R., Randazzo L. (2006), A hysteresis based scaling approach to deduce field hydraulic behaviour from core scale measurements. *Vadose Zone Journal*, 5:1005–1016 (2006), doi:10.2136/vzj2005.0128

Basile, A., G. Ciollaro, and A. Coppola, 2003. Hysteresis in soil water characteristics as a key to interpreting comparisons of laboratory and field measured hydraulic properties, *Water Resour. Res.*, 39(12), 1355, doi:10.1029/2003WR002432.

Bear J. (1979), Hydraulics and groundwater, McGraw Hill, N.Y.

Beckett P.H.T. and R. Webster (1971). Soil variability: A review, *Soil and Fertilizers*, 34:1-15.

Boels D., J.B.H.M. Van Gils, G.J. Veerman , K.E. Wit (1978), Theory and system of automatic determination of soil moisture characteristics and unsaturated hydraulic conductivities, Soil Sci., 126, 191-199.

Booltink, H.W.G., J. Bouma, D. Gimenez (1991), Suction crust infiltrometer for measuring hydraulic conductivity of unsaturated soil near saturation, *Soil Sci. Soc. Am. J.*, 55, 566-568.

Bouma J. (1983), Use of soil survey data to select measurement techniques for hydraulic conductivity, Agric. Water Managem., 3, 235-250.

Bouma J., L.W. Dekker (1978), A case study on infiltration into dry clay soil, I. Morphological observations, *Geoderma*, 20, 27-40.

Bouma, J. (1982), Measuring hydraulic conductivity soil horizons with continuous macropores, *Soil Sci. Soc. Am. J.*, 46, 438-441.

Box, G.E.P., G.M Jenkins (1970), Time series analysis, forecasting and control, *Holden day*, San Francisco.

Bresler E., H. Bielorai, A. Laufer (1979), Field test of solution models in a heterogeneous irrigated cropped soil, *Water Resour. Res.* 15, 645-652.

Brooks, R. H. and Corey, C. T. (1964), Hydraulic properties of porous media. Hydrol. Paper 3. Colorado State University, Fort Collins.

Burdine, N. T. (1953), Relative permeability calculations from pore size distribution data. Petr. Trans.. Am. Inst. Mining Metall. Eng. 198: 71-77.

Burgess T.M., R. Webster (1980a), Optimal interpolation and isarithmic mapping of soil properties, I. The semi-variogram, and punctual kriging, *J. Soil Sci.*, 31, 315-331.

Carlos M.O., J.W Hopmans, M. Alvaro, L.H Bossai, D. Wildenschild (2002), Soil water retention measurement using a combined tensiometer-coiled time domain reflectometry probe, *Soil Sci. Soc. Am. J.*, 66, 1752-1759.

Cassel D.K., O. Wendroth, D.R. Nielsen (2000), Assessing spatial variability in an agricultural experiment station field: opportunities arising from spatial dependence, *Agron. J.*, 92, 706-714.

Chen C., R.J. Wagenet (1991), Simulation of water and chemicals in macropores soils, 1. Representation of the equivalent macropore influence and its effect on soil-water flow, *J. Hydrol.*, 130, 105-126.

Chong S.K., R.E. Green, L.R. Ahuja (1981), Simple in situ determination of hydraulic conductivity by power function descriptions of drainage, *Water Resour. Res.*, 17, 1109-1114.

Ciollaro G., V. Comegna, C. Ruggiero (1987), Confronto tra metodi di campo e di laboratorio per lo studio delle caratteristiche idrauliche del suolo, Hydraulic Agriculture Insitute, University of Naples.

Comegna V., Vitale C. (1993), Space-time analysis of water status on a volcanic Vesuvian soil. Geoderma 60:135-158.

Coppola A. (2000), Unimodal and bimodal descriptions of hydraulic properties for aggregated soils, *Soil Sci. Soc. Am. J.*, 64, 1252,1262.

Coppola A., A. Comegna, G. Dragonetti, M. Dyck, A. Basile, N. Lamaddalena, M. Kassab, V. Comegna (2011), Solute transport scales in an unsaturated stony soil, *Advan. Water resour.*, 34, 747-759.

Coppola, A., Basile A., Comegna A., Lamaddalena N. (2009), Monte Carlo analysis of field water flow comparing uni- and bimodal effective hydraulic parameters for structured soil, *Journal of Contaminant Hydrology*, doi:10.1016/j.jconhyd.2008.09.007.

Coppola, A., Randazzo, L. (2006). A MathLab code for the transport of water and solutes in unsaturated soils with vegetation. Tech. Rep. Soil and Contaminant Hydrology Laboratory, Dept. DITEC University of Basilicata.

Dane J.H. (1980), Comparison of field and laboratory determined hydraulic conductivity values, *Soil Sci. Soc.Am. J.*, 44: 228-231.

David G., Geostatistical ore reserve estimation (1977), *Elsevier*, N.Y..

Day, P.R., Methods of soil analysis. Part 1. Agronomy (1965). C.A. Black Amer. Soc. of Agron., Madison, Wis. 9, 545-556.

Elrick, D.E., Scandrett, JH., E.E. Miller (1959), Tests of capillary flow scaling, *Soil Sci. Soc. Am. Proc.* 23, 329-332.

Evett, S.R., Peters F. H., Jones S. R., and Unger P. W., 1999. Soil hydraulic conductivity and retention curves from tension infiltrometer and laboratory data. Proc. Int. Workshop Characterization and Measurement of the Hydraulic Properties of Unsaturated Porous Media --University of California, pp 541-551. van Genuchten M. Th. and Leij F. J. Edt.

Frenkel H., J.O. Goertzen, J.D.R. Hoades (1978), Effects on clay type and content , exchangeable sodium percentage, and electrolyte concentration on clay dispersion and soil hydraulic conductivity, *Soil Sci. Soc. Am. J.*, 42, 32-39.

Fuentes, C., Haverkamp, R., and Parlange, J. Y. (1992), Parameter constraints on closed form soil water relationships. *J. Hydrol.* 134: 117-142.

Gajem, Y., A.W. Warrick, D.E. Myers (1981). Spatial dependance of physical properties of typic torrifluvent soil, Soil Sci. Soc. Am. J., 45, 709-715.

Gardner W.R. (1958), Some steady-state solutions of the unsaturated moisture flow equation with applications to evaporation from a water table, *Soil Sci.*, 85, 228-232.

Gerke, H.H., M.T. van Genuchten (1993), A dual-porosity model for simulating the preferential movement of water and solutes in structured porous media, *Water Resour. Res.*, 29, 305-319.

Heuvelink G.B.M., R. Webster (2001), Modeling soil variation: past, present and future, Geoderma, 100, 269-301.

Hillel, D., Environmental Soil Physics, (1998), *Academic press*, N.Y..

Hopmans J.W., J.M.H. Hendrickx, Selker J.S, (1999), Emerging measurement techniques for vadose zone characterizations. In vadose zone hydrology, Parlange and Hopmans (eds), *Oxford University Press*.

Jarvis N.J.P., P.E. Janson, P.E. Dik, I. Messing (1991), Modeling water and solute transport in macroporous soil, 1. Model description and sensivity analysis, *J. Soil Sci*, 42, 59-70.

Jensen, M.E. (1980), Design and operation of larm irrigation system, *American Society of Agricultural Engeeniers*, St. Joseph

Jones A.J., R.J. Wagenet, (1984), In situ estimation of hydraulic conductivity using simplified methods, *Water Resour. Res.*, 20, 1620-1626.

Journel A.G., C.H.J. Huijbregts (1978), Mining Geostatistic, *Accademic Press*, London.

Jury W., W.R. Gardner and W. H. Gardner (1991). Soil Physics. John Wiley and Sons Inc., ISBN 0-471-83108-5.

Kalman R.E., A new approach to linear filtering and predictions problems (1960), *Transactions of Asma, D. Jour. of Basic Enginering*, 82, 35-45.

Klute A. (ed.) (1986), Methods of soil analysis. Part 1. 2nd ed. *Agron. Monogr. 9 ASA and SSSA*, Madison, WI.

Klute A., G.E. Wilkinson (1958), Some tests of similar media concept of capillary flow: I. Reduced capillary conductivity and moisture characteristic data, *Soil Sci. Soc. Am. Proc.* 22, 278-281.

Kool, J. B., Parker, J. C., and van Genuchten, M. Th. (1987), Parameter estimation for unsaturated flow and transport models — A review. *J. Hydrol.* 91: 255-293.

Kutilek, M., and D. R. Nielsen (1994), Soil Hydrology, 370 pp., Catena, Cremlingen-Destedt, Germany.

Lax P.D. (1972), The formation and decay of shock waves. *Am. Math. Month.*, 79, 227-241, 1972.

Leij, F. J., Alves, W. J., van Genuchten, M. Th., and Williams, J. R. (1996), The UNSODA — Unsaturated Soil Hydraulic Database — User's Manual Version 1.0. Report EPA/600/R-96/095. National Risk Management Research Laboratory, Office of Research Development, U.S. Environmental Protection Agency, Cincinnati, Ohio: 1-103.

Libardi P.L., K. Reichardt, D.R. Nielsen, J.W. Bigger (1980), Simple field methods for estimating soil hydraulic conductivity, *Soil Sci. Soc. Am. J.*, 44, 3-7.

Matheron G., The theory of regionalized variables and its application (1971), *Ecole des mines de paris*. Fontainbleau, France.

Messing I., N.J. Jarvis (1993), Temporal variation in the hydraulic conductivity of a tilled clay soil as a measured by tension infiltrometers, *J. Soil Sci.*, 44, 11-24.

Miller E.E.., R.D. Miller (1955a), Theory of capillary flow: I. Practical implications, *Soil Sci. Soc. Am. Proc.*, 19, 267-271.

Miller, E.E., R.D. Miller (1955b), Theory of capillary flow: II. Experimental information, *Soil Sci. Soc. Am. Proc.* 19, 271-275.

Mishra, S. J. and Parker, J. C. (1990), On the relation between saturated hydraulic conductivity and capillary retention characteristics. *Ground Water.* 28: 775-777.

Morkoc F., J.W. Biggar, D.R. Nielsen, D.E. Rolston (1985), Analysis of soil water content and temperature using state-space approach, *Soil sci. Soc. Am. J.*, 49, 798-803.

Mualem Y., G. Dagan (1975), A dependent demain model of capillary histeresis, *Water Res. Res.*, 11, 452-460.

Mualem, Y. (1974), A conceptual model of hysteresis. *Water Resour. Res.* 10: 514-520.

Nielsen D.R., J.W. Biggar K.T. Erh (1993), Spatial variability of field-measured soil-water properties. Hilgardia. 42:215-259.

Nielsen D.R., O. Wendroth (2003), Spatial and temporal statistics. *Series: Advances in Geoecology*, Catena.

Peck A.J., R.J. Luxmoore, J.L. Stolzy (1977), Effects of spatial variability of soil hydraulic properties in water budget modeling. *Water Resour. Res.*, 13, 348-354.

Perroux K.M., I. White (1988), Designs for disc permeameters, *Soil Sci. Soc. Am. J.*, 52, 1205-1214.

Philip J.R. (1966), A linearization change for the study of infiltration, In water in the unsaturated zone, *Unesco symp.*, 471-478.

Philip J.R. (1969), Theory of infiltration, *Advances in hydroscience*, 5, 215-296.

Philip J.R. (1985), Reply to comments on `steady infiltration from spherical cavities", *Soil Sci. Soc. Am. J.*, 49, 788-789.

Philip J.R. (1986) Linearized unsteady multidimensional infiltration, *Water Resour. Res.*, 22, 1717-1727.

Poulsen T.G, P. Moldrup, O. Wendroth, D.R Nielsen (2003), Estimating Saturated hydraulic conductivity and air permeability from soil physical properties using state-space analysis, *Soil Science* 168, 311-320.

Raats P.A.C. (1976), Analytical solutions of a simplified flow equation, *Trans. Am. Soc. Agric. Eng.*, 19, 683-689.

Rao P.S.C., R.E. Jessup, A.C. Hornsby D.K. Cassel, W.A. Pollans (1983), Scaling soil microhydrologic properties of lakeland and konawa soils using similar media concepts, *Agric. Water Manage.*, 6, 277-290.

Regalado C.M., R. Munoz Carpena, A.R.M. Socorro, J.M. Hernadez Moreno (2003), Time Domain Reflectometry Models as tool to understand the electric response of volcanic soils, *Geoderma*, 117, 313-330.

Russo, D. and Bresler, E. (1980), Scaling soil hydraulic properties of a heterogeneous field. *Soil sci. Soc. Am. J.* 44:681-684.

Schuh W.M., J.W. Bauder, S.C. Gupta (1984), Evaluation of simplified methods for determining unsaturated hydraulic conductivity of layered soil, *Soil Sci. Soc. Am. J.*, 48, 730-736.

Scotter D.R., B.E. Clothier, E.R. Harper (1982), Measuring saturated hydraulic conductivity and sorptivity using twin rings.aust, *J. Soil res.* , 20, 295-304.

Sharma M.L., R.J. Luxmoore (1979), Soil spatial variability and its consequences on simulated water balance, *Water Resour. Res.*, 15:1567-1573.

Shumway R. H., D. S. Stoffer (2000), Time series analysis and its applications, *Springer Verlag*, N.Y..

Simmons C.S., D.R. Nielsen, J.W. Biggar (1979), Scaling of field-measured soil-water properties, *Hilgardia*. 47, 77-173.

Sisson J.B., A.H. Ferguson, M.Th. Van Genuchten (1980), Simple method for predicting drainage from field plot, *Soil Sci. Soc. Am. J.*, 44, 1147-1152.

Sisson J.B., M.Th. van Genuchten (1991), An improved analysis of gravity drainage experiments for estimating the unsaturated soil hydraulic functions. *Water Resour. Res.*, 27, 569-575.

Smettem K.R.J., B.E. Clothier (1989), Measuring unsaturated sorptivity and hydraulicc conductivity using multiple disc permeameters., *J. Soil Sci.*, 40, 563-568.

Snedecor G.W., W.G. Cochran (1980), Statistical methods, 7th edition, *Iowa State University Press*, Ames, Iowa.

Tamari, S., L. Bruckler, J. Halbertsma, and J. Chadoeuf (1993), A simple method for determining soil hydraulic properties in the laboratory, Soil Sci. Soc. Am. J., 57, 642– 651.

Tillotson P.M., D.R. Nielsen (1984), Scale factors in soil science, *Soil sci. Soc. Am. J.*, 48, 953-959.

Topp G.C., J.L. Davis, A.P. Annan (1980), Electromagnetic determination of soil water content: measurement in coaxial transmission lines, *Water Resour. Res.*, 16, 574-582.

Van Dam, J.C., Huygen, J., Wesseling, J.G., Feddes, R.A., Kabat, P., van Walsum, P.E.V., Groenendijk, P., van Diepen, C.A. (1997), SWAP version 2.0, Theory. Simulation of water flow, solute transport and plant growth in the Soil-Water-Atmosphere-Plant environment. Technical Document 45, DLO Winand Staring Centre, Report 71, Department Water Resources, Agricultural University, Wageningen.Ahuja L.A., J.W. Naney,D.R. Nielsen (1984), Scaling soil water properties and infiltration modelling, *Soil Sci. Soc. Am. J.*, 48, 970-973.

van Genuchten M.Th. (1980), A closed-form equation for predicting the hydraulic conductivity of unsaturated soils, *Soil Sci. Soc. Am. J.*, 44, 892-898.

van Genuchten M.Th.., D.R. Nielsen (1985), On describing and predicting the hydraulic properties of unsaturated soils, *Ann. Geophys*, 3, 615-628.

van Genuchten, M. Th., Leij, F. J., and Yates, S. R. (1991), The RETC code for quantifying the hydraulic functions of unsaturated soils, EPA/600/2-91/065, US Environmental Protection Agency, Ada, OK.

Vauclin M., G. Vachaud (1987), Caracterisation hydrodynamique des sols: analyse simplifie des essais de drainage interne, *Agronomie*, 7, 647-655.

Vauclin M., S.R. Viera, R. Bernard, J.L. Haffield (1982), Spatial variability of surface temperature along two transects of a bare soil, *Water Resour. Res.*, 18, 1677-1686.

Vieira S.R., J.L. Hatfield, D.R. Nielsen, J.W. Biggar (1983), Geostatistical theory and applications to variability of some agronomical properties, *Hilgardia*, 51, 1-72.

Warner G.S., I.D. Moore, J.L. Nieber, R.L. Geise (1989), Characterization of macropores in soil by computer tomography, *Soil Sci. Soc. Am. J.*, 53, 653-660.

Warrick A.W. (1992), Models for disc infiltrometers, *Water Resour. Res.*, 28, 1319-1327.

Warrick A.W., A. Amoozegar-fard (1979), Infiltration and drainage calculations using spatially scaled hydraulic properties, *Water Resour. Res.*, 15, 1116-1120.

Warrick A.W., G.J. Mullen, D.R. Nielsen (1977b), Scaling field-measured soil hydraulic properties using a similar media concept, *Water Resour. Res.*, 13, 355-362.

Watson K.K. (1966), An instantaneous profile method for determining the hydraulic conductivity of unsaturated porous materials, *Water Resour. Res.*, 2, 709-715.

Watson K.W., R.J. Luxmoore (1986), Estimating macroporosity in a forest watershed by use of a tension infiltrometer, *Soil Sci. Soc. Am. J.* , 50, 578-582.

Webster R. (1977), Spectral analysis of gilgai soil, *Aust. J. Soil Res*, 15, 191-204.

Webster R., C.H.E. Cuanalo De La (1975), Soil transect correlogram of north oxford shire and their interpretation, *J. Soil. Sci.*, 26, 176-194.

Wendroth O., A.M. Al-Omran, C. Kirda, K. Reichardt, D.R. Nielsen (1992), State-space approach to spatial variability of crop yield, *Soil Sci. Soc. Am. J.* 56, 801-807.

Wendroth O., S. Koszinski, E. Pena-Yewtukhiv (2006), Spatial association between soil hydraulic properties, soil texture and geoelectric resistivity, *Vadose Zone J.*, 5, 341-355.

White I., M.J. Sully (1987), Macroscopic and microscopic capillary length and time scales from field infiltration, *Water Resour. Res.*, 23, 1514-1522.

Wind, G. P. (1968), Capillary conductivity data estimated by a simple method, in Water in the Unsaturated Zone, vol. 1, edited by P. E. Rijtema and H. Wassink, Int. Assoc. Sci. Hydrol. Publ., 82– 83, 181– 191.

Wooding R.A. (1968), Steady infiltration from a shallow circular pond, *Water Resour. Res.*, 4, 1514-1522.

Wu, L., W.A. Jury, A.C. Chang, R. R. Allmaras (1997), Time series analysis of field-measured water content of a sandy soil, *Soil Sci. Soc. Am. J.*, 61, 736-742.

5

Hydrogeological-Geochemical Characteristics of Groundwater in East Banat, Pannonian Basin, Serbia

Milka M. Vidovic and Vojin B. Gordanic

The University of Belgrade, The Institute of Chemistry, Technology and Metallurgy,
Department for Ecology and Technoeconomics
Serbia

1. Introduction

Eastern Banat is the eastern part of Vojvodina Province, southern belt of the Pannonian Basin, a surface area of some 505 km² in NE Serbia (Fig. 1). It is rolling country at altitudes between 75 and 100 metres. The largest positive morphologic feature is Vrsac Mountains that extend eastward to the Romanian-Serbian border. The Mountains occupy an area of 170 km² surrounded by Mali Rit and Markovac stream in the north, Mesic and Guzijana ranges in the south, Vrsac suburbs in the west and Serbian-Romanian border in the east.

Major hydrographic features on the southern Vrsac Mountains are the streams Guzijana and Fizes by the Romanian border and the Mesic stream that drains most of the given area. The Mesic brook is 11.4 km long from the spring at el. 194 m and drains an area of 31.9 km². Highest peaks are Vrsacka Cuka (399 m), Botija Plate (469 m), Vrsacki Vrh (590 m) and Kudricki Vrh (641 m).

Major surface streams in the area are the Karas, across the area south of Vrsac Mountains, and the Nera, south of Bela Crkva. Both streams run from Romania and their discharges depend on the respective spring flows, or atmospheric precipitations over the year.

Southern Banat was prospected (Ivkovic, 1960) for the hydrogeological map of Vojvodina (Map scale 1:500,000, 1957). Background information on groundwater chemical composition in SE Banat, and Vojvodina, was reported in the sixties (Milojevic, 1963).

Most informative data on geology and production of groundwater from the aspect of data collecting and interpretation are those from Vrsac and Bela Crkva areas in eastern Banat, the SE Pannonian depression in Serbia. The Vrsac water supply source at Pavlis village was explored by drilling wells B-1 through B-17, preparing a Report with an estimate of the water supply demand by Vrsac municipality to 2005 and examining a source of low-mineral water at Mesic village (Lazic, 2010). The field works included tens drilled wells in which the water-bearing layers were deep between 27 and 110 metres, the maximum depth being about 220 metres (Vrsac). Deepest well in Bela Crkva area was 115 metres, with the water-bearing intervals from 43 to 111 metres.

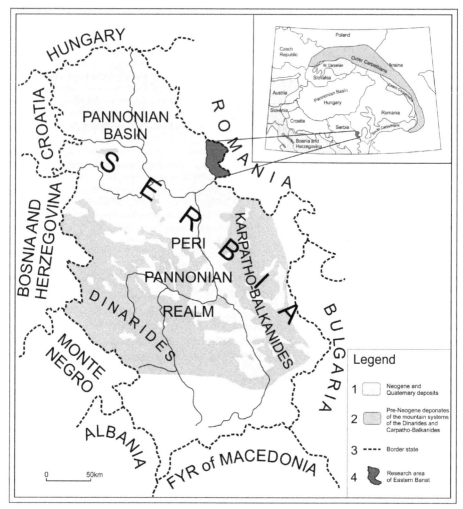

Fig. 1. Serbian part of the Pannonian Basin and the southern margin (Marovic, 2002; Dolton, 2006)

Regional investigations also yielded information about chemical composition of subsurface and surface waters (Puric & Gordanic, 1980; Cvetkovic & Gordanic, 1986; Gordanic & Jovanovic, 2007). Geochemical characteristics are particularly important for small municipal water supply systems, in view of the established elevated natural radionuclides of ^{238}U, ^{226}Ra and ^{222}Rn (on Vrsac Mountains) and other toxic elements exceeding maximum allowed concentrations (MAC) (at Bela Crkva) (Vidovic, 2007).

Extreme concentrations of U, Pb, Zn, Au and Ag, measured during the geochemical prospecting of metallic and non-metallic minerals, are correlative with the metallogenic nature of the region. There are Pb, Zn, Au and Ag mineral occurrences in eastern Vrsac Mountains in particular.

2. Methodology

Hydrogeological prospecting includes preparation of hydrogeological maps, chemical analyses of water from springs, wells and test-wells/boreholes, and continuous observation of water table fluctuation in wells, delineation of groundwater body and establishment of its type based on pumping tests.

Geochemical prospecting included sampling from different geochemical media mapped on the base geological map at scale 1:100,000 and other target maps of different scales (Fig. 2). More intensive investigation, made in Vrsac greater area for municipal water supply, was focused on lithology surveyed in boreholes (Figs. 3 and 4) and chemical analysis of groundwater (Tabs. 1 and 2).

Chemical analyses were used to classify water after Alekin (Fig. 5 a,b,c) for the Pavlis and Mesic sources (Fig. 6).

Geochemical prospecting used lithogeochemical, metallometric and hydrogeochemical techniques. For geochemical analyses samples were correlated of rocks (80 samples), stream sediments and humus (647 samples), springs (71 samples), wells (483 samples), surface streams (788 samples) and boreholes (14 samples). The results are presented on geochemical maps.

Similar results for Vrsac Mountains are given in Tab. 3.

Redeposition of natural radionuclides in the heterogeneous environment of Vrsac Mountains is expressed as $^{232}Th/^{238}U$ (Fig. 7).

The obtained concentrations of ^{238}U, ^{40}K and ^{232}Th were used to assess the effect of external equivalent of the ionizing radiation dose by Monte Carlo method (Tab. 4). Applying conversion factors for ingestion and inhalation, radionuclide intake (determined by radiometry) and the fatal cancer risk were assessed for 100,000 inhabitants for minimum and maximum activities (Tab. 5).

The parameters ^{238}U concentration, pH, Eh and Ep (Tab. 6) are determined for each water sample. Moreover, 165 samples were analysed for anion-cation composition, contents of microelements (Pb, Zn, Cu, Mo, Li, Sr, Ni and Co), of radioactive elements (U, ^{226}Ra and ^{222}Rn), of F, Br and I, dissolved oxides (SiO_2), mineral matter in water, hardness, gases CO_2, H_2S, O_2 and Rn, which were measured in the field. The excessive U, Ra and Rn concentrations in water and respective pH are given in Tab. 7 for Vrsac Mountains.

Concentrations of U, ^{226}Ra and ^{222}Rn in water (Fig. 8 a,c) are detected using two methods:

- Laser fluorometry (uranium analyzer Scintrex UA-3); the method (U-Fl-1 "Geoinstitut") measures uranium fluorescence provoked by added fluoran (inorganic phosphorous complex) that acts as a buffer for pH 7; and
- Emanometry (emanometer RDU 200 with ionization chamber) measures radium and radon in a scintillation chamber (vol. 160-170 ml) made of silver-activated ZnS. The instrument is calibrated to Bq/L unit using radium salt solution of a pre-selected concentration for three standard measures: 1, 10 and 1000 Bq/L. The obtained equivalent of 0.2489 Bq/L for 1 i/min is multiplied by the values read from the instrument.

Relative relationship of U and redox potential in water from excessive uranium areas is shown in Fig. 9, and excessive Zn, Sr, Li and Mo concentrations in relation to natural water are given in Tab. 8.

The analytical methods applied were the following:

- AAS (Atomic Absorption Spectrophotometry) for Ag, Au, Bi, Cd, Co, Cr, Cu, Mo, Na, Ni, K, Pb, Sb, V and Zn;
- AES (Atomic Emission Spectrophotometry) for Cs, Li and Rb;
- ICP-AES (ICP Atomic Emission Spectrophotometry) for B, Ba, Be, P and S.

Samples were prepared with open digestion and mineral acids or their mixtures were used for destruction. The agent used to destroy samples tested on Ag, Au, Bi, Cd, Co, Cu, K, Mo, Ni, Pb, Sb, Zn and V was aqua regia (HNO_3 : HCl). Modified aqua regia was used in destruction of the samples from which P and S were determined, and sample digestion with HF, HNO_3, $HClO_4$ to identify B, Ba, Be, Cr, Cs, Li, Rb and Sr.

Radiometric analysis for ^{238}U, ^{232}Th, ^{40}K concentrations in rocks (80 samples) used scintillation detector 4x4 Bicron with a NaJ crystal and a multiplier tube (MCA; 4096 channels) Ortec 7500 (Tab. 3, Fig. 8 b). The instrument measures flashes of a high-energy (0-3 MeV) agent. The calibration of spectra and the count of natural radionuclides, compared to the standards for uranium and thorium minerals, are NBL No. 103 0.005 % U and NBL No. 107 0.10 % Th and relative Th/U are shown in Fig. 7. The potassium standard used was potassium chloride (p.a., Merck).

3. Results and discussion

3.1 Geologic – Structure features

The geotectonic setting of Vrsac and Bela Crkva town areas is a depression formed in the early Miocene with intensive rising of the Carpathians, Dinarides and Alps (Fig. 2). The intermountain subsided and was invaded by the Mediterranean Sea. The communication with the Mediterranean Sea ceased in the late Miocene and the formed Pannonian Sea turned first into a brackish and later into a freshwater water body (Nikic & Vidovic, 2007). Neotectonic positive movements formed the horsts of Fruska Gora (a part of the Vardar Zone) and Vrsac Mountains (part of the Serbian-Macedonian Massif). The Pannonian basin has the floor pattern of minor troughs and horsts. The Vrsac Mountains and Bela Crkva sub-region is built largely of Miocene, Pliocene and Quaternary deposits with projecting Vrsac Mountains of crystalline schist and granite (Vukovic, 1965; Rakic, 1978).

Crystalline schist was found in the petroleum test wells on Vrsac Mountains at depths from 890 to 1100 metres under sedimentary rocks. The complex of crystalline rocks dated Precambrian or Lower Paleozoic underwent metamorphism during the Hercynian Orogeny. It is composed of gneiss, schist, phyllite, granite and granitic gneiss. Pegmatite, and aplite and leptynite associated with finegrained gneiss.

The Miocene (Upper Miocene, Pannonian), unconformable and transgressive over the Lower Paleozoic schist, consists of sands, clays and silts. Lower Congerian Beds or Pannonian (caspibrackish) deposits have the largest distribution east of Bela Crkva.

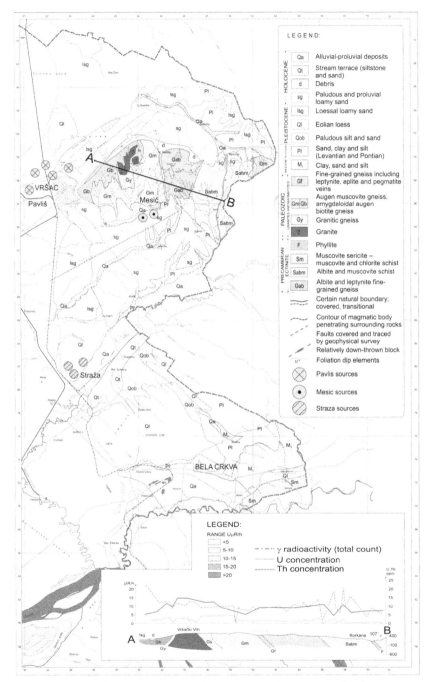

Fig. 2. Geological map of Vrsac and Bela Crkva areas (Scale 1:100,000) (Vukovic, 1965, Map compilation by Gordanic, 1980)

The Pliocene is represented by Pontian and Levantian deposits that form freshwater aquifers, which extend over the southern and central Banat and eastward into Carpathians of Romania (Vukovic, 1987). Levantian deposits with freshwater bodies are unconformable over the Pannonian, composed of gravel and sand tens of metres thick. Their lithofacial character indicates fluvial-lacustrine sediments. Lower Pliocene (Pontian) is exposed in the villages of Gudurica, Markovac, Socica, Kustilj and Jablanka near Bela Crkva, on the northern and southern Vrsac Mountains ridges and from Gudurica to Vojvodinci along the Romanian border. Minor outcrops are located at the Mesic village (IBD-1/01).

Upper Pliocene (Levantian), recognized mostly in boreholes, are gravel and sand, variable in thickness and extensive near Bela Crkva.

The Quaternary, both Pleistocene and Holocene deposits cover some 70% of the geochemically prospected Vrsac-Bela Crkva area.

Pleistocene deposits are paludous silt and sand (Qob), eolian loess (Ql) and loessal loamy sand (Lsg).

Holocene deposits form the northern and southern Vrsac foothills and extend along small and large streams (Mesic, Guzijana, Nera and Karas). Their heterogeneous composition includes paludous and proluvial loamy sand (Sg), detritus (d), stream terraces (silt and sand, Qt); alluvial-proluvial deposits (channel facies, gravel, sand and silt overbank deposits, Qa).

Two large structural units occupy the region: Rhodope, built of crystalline schist of low metamorphic grade (Rakic, 1978), and Banat or Banat-Morava unit a neoalpine structure, both extending from NE to SW, which is the main trend of the structural features (Askin, 1967) formed in result of the crystalline bedrock fracturing during the Savian and Pyrenean Orogenies.

Tectonic events of different intensities, from the Oligocene-Miocene to the present, formed horst and trough features in numerous faults. The Pavlis source area is in a deep trough that formed between the horst range Velika Greda-Jermenovci-Lokve-Nikolinci-Tilva and Vrsac Mountains, in which different chronostratigraphic units were brought to the same level. Further subsidence and diagenetic processes during the Quaternary developed a shallow surface depression, the Alibunar-Vrsac depression.

3.2 Hydrogeological characteristics

Different aquifers are related to the regional geological-structural and geolithological features of the rock masses and their porosity. Bodies of rocks that contain and conduct groundwater are either water-table or artesian aquifers.

3.2.1 Water-table aquifer

Phreatic water in the Vrsac municipality occurs in:

-   alluvial deposits from the Karas (Filipovic, 2003), Guzijana, Boruga, Markovac, Moravica (minor intermittent streams), which are heterogeneous (sand, gravelly sand, clay soil, sandy clay and mud) and have transmissibility coefficient (T) up to 22×10^{-3} m^2/s;

- eolian sand of the Deliblato Sands (Filipovic, 2003) SW of the prospected area; its filtration coefficient (K) range is from 1×10^{-4} m/s to 2×10^{-4} m/s;
- loessal deposits of 35% to 45% porosity between Potporanj-Uljma and Deliblato Sands, SW of the area.

Unconfined groundwater has been used for water supply to households on the territory of Vrsac Municipality. The depth to groundwater is between five and ten metres. The water table contours are comparable with the topographic contours, only the slope of the water table is gentler than that of the land surface. Water table contours indicate directions of the groundwater drainage to the lower lands where are the local base levels of erosion (the streams Karas and Nera and the Danube-Tisa-Danube Canal). Diffuse aquifers are fractured Paleozoic schistose rocks of Vrsac Mountains and east of Bela Crkva. The complex of schistose rocks had been much deformed by multiple folding of a structure in NE and SW directions with minor local deflections. Rocks heavily fractured at the surface have fractures and cracks mostly filled with weathering waste. The density and size of cracks decrease with the depth nearly to the point of disappearance of water table. Groundwater is recharged almost entirely from precipitations and is discharged to small morasses or at intermittent springs. Most of precipitation runs off or evaporates, with only a very small amount percolating underground.

Very weak springs of less than 0.1 L/s occur some 50 to 100 metres down the hilltops or their occurrence is secondary. The spring-discharge fluctuation is small because the capacity of aquifers is small. A change in the water-yielding capacity of the aquifer involves other modifications such as water temperature, chemical composition (amount of minerals), etc. Classified by the water-yielding capacity of aquifers, this area is assigned to the low-yield group from 0.1 to 1.0 L/s per discharge point.

Surface streams that are draining the area run from NE to SW, same as the trends of the structural features. Groundwater flows through the weathering crust mainly in the same direction towards the local base level of erosion. Unconfined water does not occur below the base level, not even in fractures intersecting the area.

Groundwater is dominantly HCO_3-Ca-Mg, low mineralized (0.4 to 0.5 g/L), neutral (pH 7), contained in an unconfined hydrogeologic body of intensive water exchange, so that individual chemical constituents are susceptible to transformation with the time.

3.2.2 Artesian aquifer

A Neogene complex of Miocene lithological varieties more than a thousand metres thick forms an artesian water body between Miocene or Paleozoic crystalline schist and intrusive granite, where Miocene deposits are lacking, and 30-100 metres thick Quaternary.

Dense-rock aquifers are mostly Tertiary or Quaternary intergranular sand deposits, like in borehole IBD-1/0 at Mesic village (Fig. 3), either water-table (shallow, phreatic) or confined (deep, artesian or subartesian) aquifers.

A shallower artesian aquifer, less than 200 metres deep, is formed by Upper Pliocene water-bearing rocks that are not clearly differentiated from other rocks, because water-bearing layers occur at different depth intervals in boreholes within short distances.

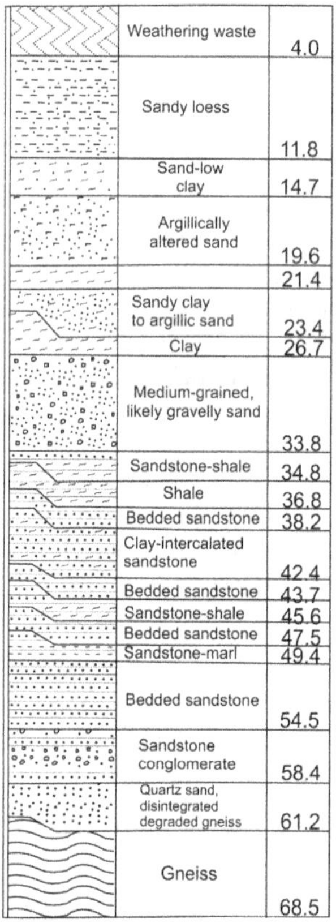

	Weathering waste	4.0
	Sandy loess	11.8
	Sand-low clay	14.7
	Argillically altered sand	19.6
		21.4
	Sandy clay to argillic sand	23.4
	Clay	26.7
	Medium-grained, likely gravelly sand	33.8
	Sandstone-shale	34.8
	Shale	36.8
	Bedded sandstone	38.2
	Clay-intercalated sandstone	42.4
	Bedded sandstone	43.7
	Sandstone-shale	45.6
	Bedded sandstone	47.5
	Sandstone-marl	49.4
	Bedded sandstone	54.5
	Sandstone conglomerate	58.4
	Quartz sand, disintegrated degraded gneiss	61.2
	Gneiss	68.5

Fig. 3. Columnar section for borehole IBD-1/01 in the Mesic village (Lazic, 2002)

The groundwater body is difficult to outline, because it extends into Romania on one side and throughout the Pannonian basin on the other side. Its southern boundary is the Danube, and the eastern boundary is Vrsac Mountains (the SW offsets of the Carpathians) and a crystalline massif near Bela Crkva. Characteristic geological and hydrogeological sections beneath Quaternary deposits in Vrsac area near the artificial lake are reconstructed on the data from RB-1 and B-1 test-wells. Lithology of Paludine-Levantian sands under Quaternary deposits consists of the following intervals: 0-23 m Quaternary complex (clay, gravel, sand); 23-42 m grey-green sand; 42-44 m clayey sand; 44-67 m grey-green fine- to medium-grained sand on gravel; 67-73 m grey-green silty to fine sand; 73-96 m grey clay; 96-121 m grey medium-grained sand; 121-174 m grey clay; 174-188 m grey coarse quartz sand; and 188-220 m grey clay. The depth of good hydraulic properties is between 23 and 67 metres, particularly interval 44-67 m, confirmed by pumping test in well RB-1 (Milivojevic, 2004).

The hydrogeological setting is indicated by drilling data from B-1 (Gordanic, 1980) (Fig. 4).

On the profile of the borehole (Fig. 4), there was found four water-bearing horizons, which are mutually hydrogeological different. Deeper than these lithological members, there are water-bearing layers, which are not included from the point of these studies.

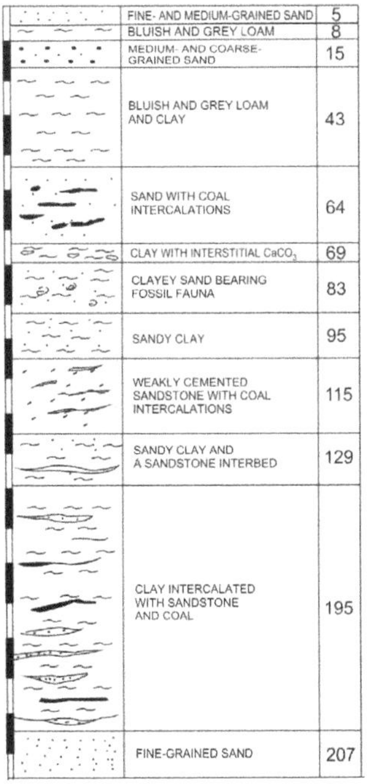

	FINE- AND MEDIUM-GRAINED SAND	5
	BLUISH AND GREY LOAM	8
	MEDIUM- AND COARSE-GRAINED SAND	15
	BLUISH AND GREY LOAM AND CLAY	43
	SAND WITH COAL INTERCALATIONS	64
	CLAY WITH INTERSTITIAL $CaCO_3$	69
	CLAYEY SAND BEARING FOSSIL FAUNA	83
	SANDY CLAY	95
	WEAKLY CEMENTED SANDSTONE WITH COAL INTERCALATIONS	115
	SANDY CLAY AND A SANDSTONE INTERBED	129
	CLAY INTERCALATED WITH SANDSTONE AND COAL	195
	FINE-GRAINED SAND	207

Fig. 4. Columnar section for borehole at the Vrsac artificial lake (Gordanic, 1980)

The Vrsac borehole (Fig. 2) penetrated four different water-bearing layers: Upper Pontian, Levantian and Pleistocene clay-sand deposits form several confined (artesian and subartesian) aquifers to a depth of 300-400 metres from the surface. Mineral salts and temperature of water are increasing with the depth, yet this is the main source of domestic and industrial water supply. Moreover, the prevailing hydraulic system is complex. Strata of deposits that include confined water bodies are interrelated because of the frequent vertical and lateral variations in facies. Horizontal and vertical communication of confined groundwater has also hydraulic connection with the shallow unconfined groundwater. Artesian water-bearing layers, established on the Vrsac Mountains margin, dip to the central basin along fault zones, which provides for overflow of mineral-high water from deep into the shallow layers. The test hole near the artificial lake at Vrsac produced water from the depth of 200 metres that had mineral content 6.91 g/L and temperature of 24°C. Groundwater from the same depth in Banat, only unaffected by overflow from deeper aquifers commonly contains 1.1 to 1.6 g/L dissolved mineral material. Water from deep aquifers in Vrsac area moves along the contact of Tertiary deposits and crystalline schists and overflows into the shallower Pliocene aquifers of lower confining pressure.

The main groundwater reservoir (both artesian and subartesian) is replenished directly by percolating atmospheric precipitations on Vrsac Mountains margin and Bela Crkva area, and in Romanian side by infiltration from rivers. Indirect replenishment by precipitations occurs from water-table aquifer in the Deliblato Sands and on the loess plateau. Groundwater recharge during piestic low of the main reservoir in the Alibunar-Vrsac depression is small, indirect from the overlying semipermeable layer and from drainage canals.

Groundwater circulation is slow, decreasing from the Neogene basin borderland where water ^{14}C dated was 1100 years old (Gudurica). Further from the edge of the basin, the water age increases to 6950±160 years in Potporanj and 9900±100 years at Pavlis.

Groundwater in the main reservoir is chemically characterized by low TDS, with dry residue 280-420 mg/L, locally about 600 mg/L, pH 7.4-7.6 units. Total hardness varies from 12 to 16.5 °dH. The KMnO$_4$ rate is low, not exceeding 5 mg/L. Ammonia is always higher (0.6-1.2 mg/L) than MAC, and local concentrations of iron and manganese are 2.2 and up to 0.15 mg/L, respectively. Other mineral constituents are within the standards for drinking water.

3.3 Hydrochemical exploration of main groundwater reservoir

Results of the hydrochemical exploration are presented for the main reservoir in the Pavlis source and aquifers in the Mesic areas.

3.3.1 Pavlis source area

Groundwater in Pavlis area is used for municipal water supply to Vrsac and to small communities in the surroundings. Water production started in 1961 in an area bordered by the Keveris and Veliki Kanal and the village of Pavlis. A Feasibility Study was prepared in 1987, a water supply projection to 2005 needed by the growing town and its industry, from the main groundwater reservoir. On the basis of the Study, the source area has been extended to the present 17 wells (B-1 to B-17).

The production of 10 L/s at first increased to 180 L/s. When the Study was prepared, a drawdown of about 0.5 m/yr was recorded in the Pavlis area. The exploitation regime at Pavlis is considered to be "relaxed", about 16 hours per day on average. Chemical composition of water was analysed previously in relation to the regime of production. Before 1977, only wells B-7 and B-8 in the main reservoir were analysed. After 1977, seven new wells were drilled from which natural water from the main aquifer has been analysed. The parameters analysed are: Na$^+$, K$^+$, CaO, Ca^{2+}, MgO, Mg^{2+}, Cl$^-$, SO$_4^{2-}$, HCO$_3^-$, Fe, Mn, NO$_3^-$, NO$_2^-$, NH$_3$, F, SiO$_2$, pH, hardness, Ep, KMnO$_4$ consumption and M (mineralisation). Water quality deterioration has not been registered for any parameter in the given period.

Chemical composition of groundwater in Pavlis source area is expressed by Kurlov formula (1) (Slimak, 2005) as:

$$M_{0.284}\frac{HCO_{97}^3 Cl_3}{Ca_{42}Mg_{31}Na + K_{27}} = t_{16.7} \tag{1}$$

and Table 1 gives the quality parameters resulting from chemical analyses for wells B-17, B-6 and B-7 in Pavlis source.

The amount of oxidizing organic substances is expressed by KMnO$_4$ consumption, the highest recorded being 3.1 mg/L for well B-7. The critical inorganic parameters are ammonia ion and Mn, which are excessive in each sample. Concentrations of heavy metals are at the detection level for the applied method.

PARAMETER	Test Results		
	B-17 1 July 2004	B-6 1 Nov. 2004	B-7 31 Mar. 2005
Colour, °Pt-Co	-	-	< 5
Odour and taste	-	-	+
Turbidity, NTU	< 0.6	< 0.6	0.55
pH	7.4	7.7	7.6
Oxidizability, mg $KMnO_4/L$	0.7	1.8	3.1
Temperature, °C	16.5	16.8	15.7
Dry residue, mg/L	312	417	430
Electrical conductivity, µS/cm	494	624	703
Hardness, °dH	12.2	17.5	18
Sulphates, mg/L	< 1	< 1	0.7
Ammonia (NH_4), mg/L	1.5	1.7	1.58
Arsenic (As), mg/L	< 0.005	< 0.005	< 0.004
Copper (Cu), mg L	< 0.002	< 0.002	< 0.01
Barium (Ba), mg/L	< 0.4	< 0.4	
Bromides (Br), mg/L	< 0.01	< 0.01	
Cyanides (CN), mg/L	< 0.002	< 0.002	< 0.01
Zinc (Zn), mg/L	< 0.005	< 0.005	< 0.01
Fluorides (F), mg/L	< 0.2	0.24	0.24
Phosphates (P), mg/L	< 0.03	< 0.03	0.01
Chromium total (Cr), mg/L	< 0.005	< 0.005	< 0.001
Chlorides (Cl), mg/L	9.7	14.2	20
Hydrocarbonates, mg/L	518	378	482
Cadmium (Cd), mg/L	< 0.002	< 0.002	< 0.001
Calcium (Ca), mg/L	85.4	54.5	72.5
Potassium (K), mg/L	9.3	5.6	2.1
Cobalt (Co), mg/L	< 0.005	< 0.005	
Lithium (Li), mg/L	< 0.01	< 0.01	
Magnesium (Mg), mg/L	40.1	29.7	34.3
Manganese (Mn), mg/L	0.15	0.21	0.11
Sodium (Na), mg/L	38.5	33.5	47.4
Nickel (Ni), mg/L	< 0.005	< 0.005	< 0.01
Nitrates (NO_3), mg/L	< 0.5	< 0.5	< 0.5
Nitrites (NO_2), mg/L	< 0.005	< 0.005	< 0.005
Lead (Pb), mg/L	< 0.005	< 0.005	< 0.01
Silicates (SiO_2), mg/L	35.0	34.1	28.18
Silver, mg/L	< 0.005	< 0.005	
Strontium, mg/L	< 0.05	< 0.05	
Mercury (Hg), mg/L			< 0.001
Aluminium, mg/L	< 0.01	< 0.01	< 0.05
Iron, mg/L	0.2	0.15	0.04

Table 1. Hydrochemical analysis of groundwater for Pavlis source

In Alekin classification, groundwater from the Pavlis main reservoir is hydrocarbonate-calcium-magnesium water, its chemistry in the reported period (Fig. 5 a, b and c).

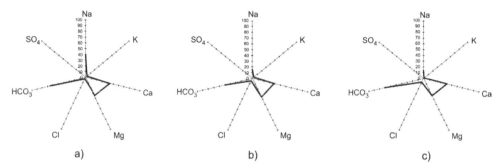

Fig. 5. Groundwater from Pavlis source in Alekin classification. a) Well B-17 on 1 July 2004; b) Well B-6 on 1 Nov. 2004; and c) well B-7 on 31 Mar. 2005

3.3.2 Mesic source area

Artesian aquifer (the village of Mesic) is located on the southern side of Vrsac Mountains. Water is pumped from three rock units: crystalline schist, gneiss, Neogene deposits and Quaternary alluvial deposits of the Mesic brook.

A structural feature in the area is a fault with the northern block of some 20 metres relative downthrow. The Mesic Brook runs from the source in southward direction from the fault, then deviates 90° to SW. In context of local geology and hydrogeology, groundwater reservoirs must have formed in an open structure, and in semi-open Pleistocene quartz sand. Based on the geological setting, porosity of lithostratigraphic units and hydrodynamic conditions, the types of groundwater bodies (Lazic, 2010) are the following:

- Compact water-table aquifer of considerable water-yielding capacity in alluvial deposits of gravel and sand;
- Compact subartesian aquifer of smaller water-yielding capacity in Pontian and Pleistocene sand deposits (IBD-1/01 Mesic); and
- Fractured-rock aquifer of low water-yielding capacity in muscovite gneiss.

Table 2 gives chemical data of groundwater from test/production well IBD-1/01 in a low-capacity aquifer, for period 2008-2009. The water-bearing formation is largely quartz sand in two layers (lower/basal and higher/upper) at different depths, whose total thickness in test/production well IBD-1/01 is some 10 metres.

The main water-bearing formation consists of two (lower and upper) layers of about 10 m total thickness in test well IBD-1/01 area of influence. The formation is overlain by Quaternary clay and sand 10-15 m thick (26.7 m in the well-influence area) and underlain by crystalline schist. Loam is a kind of confining bed for the surface Mesic brook surface water. Groundwater in this formation is under mild subartesian pressure from the mentioned upper layer.

Pressure in the lower layer of quartz sand is somewhat higher, to 1.5 m below the surface, although IBD-1/01 is 2-3 m higher than BD-1/02. This water-bearing layer, where sands under a thin cover lie near the surface, is replenished by atmospheric precipitations that infiltrate from the crushed and fractured contact zone with crystalline schist.

PARAMETER	Well IBD-1/01 test results			
	25 Mar. 08	26 June 08	13 Apr. 09	12 Oct. 09
Temperature, °C	-	15.2	15	15.3
Colour	clear	< 5	< 5	< 5
Turbidity, NTU	< 0.6	0.1	0.2	0.2
pH	6.3	6.4	6.3	6.2
$KMnO_4$ consumption, mg $KMnO_4/L$	1.3	0.3	0.4	1.2
Dry residue at 180°C, g/L	0.185	0.177	0.17	0.191
Electrical conductivity, $\mu S/cm$	247	250	250	250
Free hydrogen sulphide (H_2S), mg/L	< 0.02	< 0.02	-	< 0.02
Free carbon dioxide (CO_2), mg/L	-	69	82.8	99.7
Total hardness, °dH	6.70	6.30	6.10	6.10
Hydrocarbonates (HCO_3), mg/L	134	130.5	122	129.3
Nitrites (NO_2), mg/L	< 0.004	< 0.006	< 0.006	< 0.006
Nitrates (NO_3), mg/L	15.7	16.2	17.5	18.1
Chlorides (Cl⁻), mg/L	9.3	10.4	9.2	10.4
Sulphates (SO_4^{2-}), mg/L	14.4	11.9	12.3	15.2
Ortho-phosphates(P), mg/L	< 0.06	< 0.02	0.03	< 0.02
Fluorides (F), mg/L	< 0.1	0.022	< 0.05	0.09
Phenols, mg/L	< 0.002	< 0.001	< 0.001	< 0.001
Arsenic (As), mg/L	< 0.005	< 0.001	< 0.001	< 0.001
Copper (Cu), mg/L	< 0.05	0.002	< 0.002	0.003
Zinc (Zn), mg/L	< 0.005	0.0014	0.004	0.009
Iron (Fe^{++}), mg/L	0.05	0.027	0.001	< 0.004
Total chromium (Cr), mg/L	< 0.002	< 0.001	< 0.002	< 0.002
Cadmium (Cd), mg/L	< 0.001	< 0.0005	< 0.0008	< 0.0008
Calcium (Ca^{2+}), mg/L	28	31.7	30.8	30.7
Potassium (K^+), mg/L	1.1	1.08	1.14	1.19
Magnesium (Mg^{2+}), mg/L	6.8	8	7.6	7.89
Manganese (Mn), mg/L	< 0.01	0.0006	< 0.0002	< 0.0002
Sodium (Na^+), mg/L	5.6	11.3	10.7	11
Nickel (Ni), mg/L	< 0.001	< 0.004	< 0.006	< 0.006
Lead (Pb), mg/L	< 0.005	< 0.007	< 0.005	< 0.005
Mercury (Hg), mg/L	-	< 0.0005	< 0.0005	< 0.0005

Table 2. Hydrochemical analysis for well IBD-1/01 in the Mesic source area

Groundwater is assigned on the basis of chemical results (Tab. 2) to the group of hydrocarbonate-calcium-magnesium water of the Alekin classification. In cationic composition, calcium ions (28-31.7 mg/L) are dominant, followed by 6.8-8 mg/L Mg, then 5.6-11.3 mg Na^+/L and 1.08-1.19 mg K^+/L. In anionic composition, hydrocarbonate ions are dominant in concentrations from 122 to 134 mg/L. Sulphate ions are 11.9 to 15.2 mg SO_4^{2-}/L, and chloride ions are 9.2-10.4 mg Cl⁻/L.

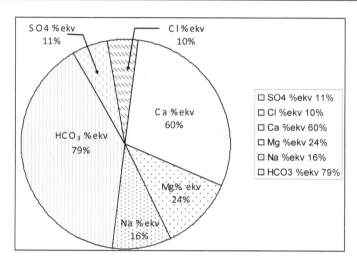

Fig. 6. Circle graph of the principal anion and cation mean concentrations in water from IBD-1/01 at Mesic village in the observation period (from March 2008 to October 2009)

Stability of chemical macro-composition or of individual constituents is based on the proportion of minimum and maximum concentrations of some elements expressed by the stability coefficient (K) for the period from March 2008 to October 2009.

The coefficient of stability is determined only for cations and anions and may be expressed by Kurlov formula. Its value does not exceed 5%, viz.: 6.1056% $K_{Ca^{2+}}$, 7.9234% $K_{Mg^{2+}}$, 29.534% K_{Na^+}, 4.653% K_{HCO3^-}, 12.268% $K_{SO4^{2-}}$ and 6.1069% K_{Cl^-}. These coefficients indicate stable chemical composition of water in the 'compact low-capacity aquifer' drained by well IBD-1/01. This refers to the dominant anions and cations (HCO_3^-, Ca^{2+} and Mg^{2+}) which qualify the type of water. The type of water from IBD-1/01 is hydrocarbonate-calcium-magnesium, written as (Kurlov formula):

$$M_{0.18} \frac{HCO_{79}^3 SO_{11}^4 Cl_{10}}{Ca_{60} Mg_{24} Na_{16}} = t_{15.15} Q_{0.9} \qquad (2)$$

In relation to the pumping coefficient ($Q_{0.9}$), the water resource is classified as "economic" or "B reserve" (Lazic, 2010).

3.4 Geochemical exploration

Geological-radiometric profiling and different geochemical prospecting techniques, adjusted to the given geological-structural setting, indicated areas of extreme metallic mineral and natural radionuclide concentrations in Vrsac Mountains. Two zones of elevated U, Mo, Pb (1-360) ppm), Zn (1-1800 ppm), Ag (1-48 ppm), Au (0.02-0.24 ppm) were detected in gneiss in the eastern part of the area (Gordanic, 1980).

Occurrences of deposits and veins of hydrothermal quartz with pyrite, magnetite and chalcopyrite in salbands at the contact with schist were located south of Gudurica in Vrsac

Mountains. Geochemical anomalies for W (3000-5000 ppm) and Mo (up to 250 ppm) were detected in granitic gneiss near a granite intrusion, south of Vrsac peak (Botija Plate). Concentrations of ^{238}U, ^{232}Th, ^{40}K, total uranium (Ut) and dissolved uranium (Ud) are determined (Tab. 1) in samples collected during the radiometric profiling (Gordanic, 1980; Spasic-Jokic & Gordanic, 2010a).

	GRANITE AUGEN BIOTITE GNEISS (Gb) GRANITE - GNEISS (Gy) i AUGEN MUSCOVITE GNEISS (Gm)			ECTINITE ALBITE LEPTYNOLITE GNEISS (Gab) ALBITE MUSCOVITE SCHIST (Sabm)		
	min	max	mean	min	max	mean
^{238}U, ppm	0.58	5.57	1.81	0.25	4.41	1.78
^{232}Th, ppm	0.22	22.2	7.66	0.84	22.08	9.75
^{40}K, %	1.82	5.39	3.75	0.27	9.92	1.75
Ut, ppm	0.2	22.0	1.31	0.1	7.6	0.55
Ud, ppm	0.1	21.0	0.71	0.1	2.1	0.40

Table 3. Radiometric results for granite-migmatite and ectinite-gneiss (Gordanic, 1980)

Natural radionuclides have a log-normal distribution in the heterogeneous environment of the Vrsac Mountains. Concentrations of uranium in granite-migmatite indicate its redistribution in the older metamorphic rocks during a long geological period, or it resulted from subsequent hydrothermal and metasomatic processes. Dominant character of Th is represented in relation to U on a Th/U diagram in Fig. 7.

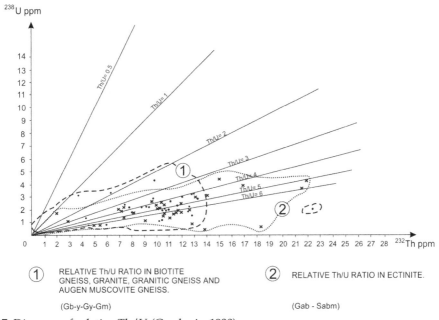

① RELATIVE Th/U RATIO IN BIOTITE
GNEISS, GRANITE, GRANITIC GNEISS AND
AUGEN MUSCOVITE GNEISS.

(Gb-y-Gy-Gm)

② RELATIVE Th/U RATIO IN ECTINITE.

(Gab - Sabm)

Fig. 7. Diagram of relative Th/U (Gordanic, 1980)

Potassium is twice higher in gneiss than in ectinite, a likely consequence of metamorphism with the supply of magmatic melts and fluids during the Hercynian Orogeny.

The effect of natural nuclides on rural environment was assessed on the Vrsac Mountains using the Monte Carlo method (Spasic-Jokic & Gordanic, 2010a) to establish external equivalent of the ionizing radiation dose 1 m above the ground, which depicts the natural environmental radiation; this in view of the natural uranium concentrations of 99.284% ^{238}U + 0.711% ^{235}U + 0.0058% ^{234}U and negligible other isotopes. Annual effective doses from ingestion (UNSCEAR, 2000) are given in Tab. 4.

INGESTION	Average (UNSCEAR, 2000)	Bela Crkva and Vrsac (present study)
^{40} K, µSv	0.17	0.004
U and Th series, µSv	0.12	0.017
Total ingestion exposure, µSv	0.29	0.021

Table 4. Annual effective dose from ingestion of terrestrial radionuclides

The effective doses for ingestion in Table 4 are lower than those of UNSCEAR.

Applying conversion factors for ingestion and inhalation, radionuclide intake (determined by radiometry) and the fatal cancer risk were assessed per 100,000 inhabitants for minimum and maximum activities (Zupunski et al., 2010; Spasic-Jokic & Gordanic, 2010b) (Tab. 5).

SOURCE	ACTIVITY (Bq)		RISK INHALATION		RISK INGESTION		Normalized on 100,000 Inhabitants	
	mean	range	mean	range	mean	range	mean	range
^{238}U	13.5	1.60-36.0	0.9×10^{-5}	$(0.1\text{-}2.3) \times 10^{-5}$	47.1×10^{-8}	$(5.7\text{-}126) \times 10^{-8}$	0.9	0.1-2.4
^{232}Th	21.1	1.70-45.2	2.3×10^{-5}	$(0.2\text{-}5) \times 10^{-5}$	6.10×10^{-8}	$(0.40\text{-}11) \times 10^{-8}$	2.3	0.2-5
^{226} Ra	12.8	1.60-34.8	0.8×10^{-5}	$(0.1\text{-}2.3) \times 10^{-5}$	44.9×10^{-8}	$(5.5\text{-}122) \times 10^{-8}$	0.9	0.1-2.4
^{40}K	490	42.4-1560	0.3×10^{-5}	$(0.02\text{-}0.8) \times 10^{-5}$	28.3×10^{-8}	$(2.5\text{-}93) \times 10^{-8}$	0.3	0.02-0.9

Table 5. Absolute fatal cancer risk assessments after exposure to radionuclides ^{238}U, ^{232}Th, ^{226}Ra and ^{40}K

The triple higher Ut concentration in granite and gneiss than in ectinite suggests a primary source of uranium. Favourable pH, Eh, Ep (Table 6) must have provided for uranium leaching from adjacent rocks, migration and precipitation in suitable geochemical environments.

Parameter	Min	Max
U, ppb	0.1	166
pH vrednost	5.5	7.9
Eh, mV	- 65	+ 190
Ep, mSv	160	4545

Table 6. Water analysis during the regional hydrogeochemical prospecting

The presence of augen biotite gneiss (Gb) around the granite intrusion on Vrsac Mountains, then the presence of granite varieties, largely altered by sericitization, muscovitization and feldspathization a younger phase of which is expressed through albitization, are the reasons for elevated U, ^{226}Ra and ^{222}Rn concentrations in water of the following anomaly zones:

- Socica-Korkana-Markovac,
- Mesic-Malo Srediste and
- Vrsacka Kula-Mali Rit.

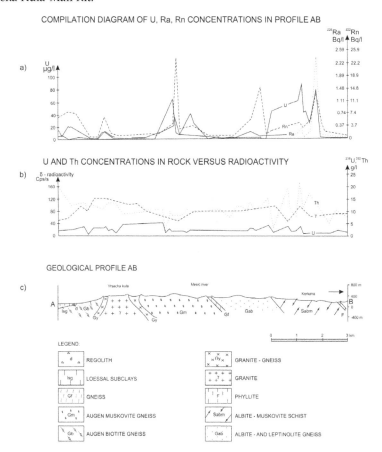

Fig. 8. a) Compilation diagram of extreme concentrations of uranium in water and b) U and Th in rocks related to the c) geologic-structure setting (Gordanic & Zunic, 1994)

The exploration data were used to delineate anomaly zones, for which geological section A-B is drawn Fig. 2 that shows the proportions of radionuclides in water and rocks. Proportions of U, Ra and Rn concentrations in water of the anomaly zones, radioactivity and U and Th concentrations in rocks are diagrammatically represented in geologic section AB, Fig. 8 a, b, c (Gordanic & Zunic, 1994).

Concentrations of U, Ra and Rn in well and spring waters of Vrsac and Bela Crkva areas are given in Tab. 7.

Classified by anionic and cationic composition (Kurlov formula), depending on the quality, waters in the exploration area range:

$$\text{from } \frac{HCO_3 - SO_4 - Cl - NO_3}{Ca - Mg - Na + K} \text{ to } \frac{SO_4 - HCO_3 - Cl - NO_3}{Na + K - Mg - Ca}.$$

Parameter	Range	Mean
U, ppb	0.1 - 166	4.2
^{226}Ra, Bq/L	0.02 - 5.92	0.11
^{222}Rn, Bq/L	0.74 – 3.33	2.74
pH	5.9 - 7.9	7.2

Table 7. Concentrations of U, Ra and Rn in waters and pH for Vrsac and Bela Crkva areas

Waters with excessive concentration of uranium have pH from 6.5 to 7.6. The range of dissolved minerals in water is from 138 mg/L to 4557 mg/L. Uranium-high waters are mainly low in dissolved mineral materials (< 1 g/L), only few are > 1 g/L to 5 g/L, suitable for uranium migration in an aquatic environment. The amount of SiO_2 in water of the anomaly zones is about 30 mg/L, which is related to the weathering zone, and somewhat higher (up to 99 mg/L) in water in contact with structural features.

The oxidation-reduction potential (Eh), determined in each water sample, varies within the range from -65 mV to +190 mV. The diagram of relative Eh and uranium relationship (Fig. 9) shows that waters in extreme zones are dominantly oxidizing, with Eh from 130 mV to 170 mV. Such an environment enhances uranium solution from adjacent rocks, which is manifested in the extreme concentration zones. In view of the metallogenic character of the explored area, waters were analyzed on concentrations of the microelements Zn, Pb, Cu, Sr, Li, Mo, Ni, Co, F, Br, I, Fe^{2+}, Fe^{3+} and SiO_2. Elevated concentrations of Zn, Sr, Li and Mo can be used as direct indication of concealed ore bodies and of uranium mineral (Tab. 8). Other analyzed elements were found in ore or less standard amounts for natural waters.

Elevated Li, Sr, U, Ra and Rn are related to the nearness of a granite intrusion, a zone of fine-grained gneiss with leptynite, aplite and pegmatite veins. The presence of Li in subsurface and surface waters may be an indication of pegmatite deposits in respect of extreme U, Mo, Pb, Zn and Ag contents in gneisses (Hawkes & Webb, 1960), which are potential sources of elevated concentrations of radioactive elements in waters.

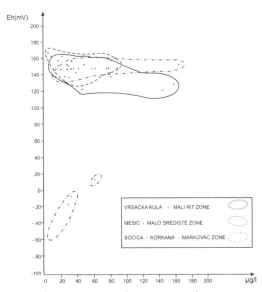

Fig. 9. Relative relationship of redox potentials in waters of extreme uranium concentration zones (Gordanic, 1980)

	min	max	Mean	NATURAL WATER, (Hawkes & Webb, 1960)
Zn, ppb	3	3800	13	1 – 200
Sr, ppb	30	3250	300	–
Li, ppb	3	60	5	0.3 – 3
Mo, ppb	1	14	2	0.05 – 3

Table 8. Amounts of Zn, Sr, Li and Mo in water

Unconfined groundwater in Upper Pliocene (Levantian) deposits of Bela Crkva is pumped for domestic water supply in small communities. Water in most wells (3-34 m deep) is below the MAC standard for drinking water (Drinking Water Regulations, Official Gazette SRY No. 42/98 and 44/99) being highly mineralized, with elevated amounts of nitrogen compounds, high conductivity and concentrations of heavy metals: Pb, Cd, Ni, Br, Al (Vidovic, 2007). Anionic and cationic compositions of waters are dominantly HCO_3-SO_4-Cl to HCO_3-Cl-SO_4, and Ca-Mg to Ca-Na, respectively. Particularly elevated are the amounts of NH_4, NO_3, NO_2, SO_4, K, Ca and F (in Kustilj and other minor communities). Water in all wells is oxygen-high between 3 and 13.1 mg/L. The effect of human activities is manifested at the water-table level. Large communities, such as Bela Crkva, supply water from drilled wells in which the depth to water table varies from 43 to 111 metres.

4. Conclusion

Controlled by the geological-structural setting of the region, groundwater bodies have formed on Pontian-Levantian deposits that extend over the southern, central and eastern

Banat and further to Carpathian offsets in Romania. Zones of groundwater recharge are the following:

- Deliblato Sands, percolation of precipitation;
- Vrsac Mountains offsets (in water-bearing layer exposures), and
- Carpathians offsets in Romania (in aquifer exposures).

Groundwater drainage in the Vrsac-Bela Crkva area operates through groundwater bodies in:

- overbank deposits from the Nera and Karas streams, or
- rises through the overlying less permeable layer to the shallower aquifer in a large confined area of groundwater (piezometric surface above the land surface),
- through many flowing wells and
- water-intake facilities.

The zones of extreme natural U, Ra and Rn radionuclides are related to the geological-structural and metallogenic features of Vrsac Mountains and have no effect on the Pavlis groundwater source for supply to Vrsac and its suburbs.

The identified extreme concentrations of Pb, Zn, Mo, Ag, Au and W in Vrsac Mountains have no effect on groundwater in Pavlis area.

Total concentrations of ΣU, ^{232}Th and ^{40}K are used to assess the fatal cancer risk per 100,000 inhabitants exposed to ingestion and inhalation of natural radionuclides in soil. Measured concentrations UNSCEAR, 2000) in the region of Vrsac Mountains are within the allowed range. This information is important for the geomedical status of the region and for other environmental-geochemical research.

Extreme concentrations (above MAC) of Pb, Li, Sr, Mo in groundwater of Vrsac Mountains are noteworthy because this water is used in rural households (Drinking Water Regulations, Official Gazette SRY No. 42/98 and 44/99). Water from tested wells used in the communities near Bela Crkva does not meet the standard for drinking water. The above-mentioned elements are below the MAC in the source of the Bela Crkva water supply. No impact of human activities in the nearby communities has been established on the source of Bela Crkva water supply.

5. Acknowledgment

This work was supported by the Ministry of Education and Science (Project OI 176018).

6. References

Askin, V. (1967). *Theoretical Classification Problems of Petroleum and Gas Deposits and Its Importance for Exploration with Particular Reference to Banat* (in Serbian). Special Edition, Vol.15, Institute for Geological and geophysical exploration, Belgrade.

Cvetkovic, D. & Gordanic, V. (1986). 1986 Report on Basic Uranium Prospecting in Socica-Markovac, Vrsac Mountains (in Serbian). Documentation fund, Geoinstitut, Belgrade.

Dolton, G.L. (2006). Panonian Basin Province, Central Europe (Province 4808) – Petroleum Geology, Total Petroleum Systems, and Petroleum Resource Assessment: U.S. Geological Survey Bulletin 2204-B, 47 p.

Filipovic, B. (2003). *Mineral, Thermal and Thermomineral Waters of Serbia. Monograph* (in Serbian). Udruzenje banjskih i klimatskih mesta Srbije, Vrnjacka Banja. Faculty of Mining and Geology, Institute of hydrogeology, Belgrade.

Gordanic, V. (1980). Special Geochemical Part, In: *Report on Regional Exploration of Nuclear Mineral Resources in Vrsac Hills and Bela Crkva Area* (in Serbian), Puric, D. & Gordanic, V., 1980, Geoinstitut, Belgrade.

Gordanic, V. & Zunic, Z. (1994). Geochemical Prospecting, in: *1st Mediterranean Congress on Radiation Prospecting, Proceedings*, pp. 291-294. Athens 1994, Greece.

Hawkes, H.E. & J.S. Webb (1960). Geochemistry in Exploration of Mineral Resources (in Serbian), Belgrade.

Ivkovic, A. (1960). Hydrogeology of South Banat (in Serbian). Documentation fund, Geological Institute, Belgrade.

Lazic, M. (2002). Report on the Low-Mineral Water Reserve in Mesic- Vrsac Municipality (in Serbian). Documentation fund, Municipality of Vrsac.

Lazic, M. (2010). Revised Report of Groundwater Reserve at Mesic for the Moja Voda Bottling Plant.(in Serbian). Documentation fund, Faculty of Mining and Geology, Hydrogeology Department.

Marovic, M., Đokovic, I., Pesic, L., Radovanovic, S., Toljic, M. & Gerzina, N. (2002). Neotectonics and Seismicity of the Southern Margin of the Pannonian Basin in Serbia. *EGU Stephen Mueller Special Publications*, Series 3, pp. 277-295.

Milivojevic, M. (2004). A Study: Quality of Water and Sediment in the Vrsac Lake and Assessment of the Groundwater Resource in the Surrounding Area (in Serbian). Documentation fund, Faculty of Mining and Geology, Institute of hydrogeology, Belgrade.

Milojevic, N. (1963). Hydrochemical Character and Regime of Groundwaters in Vojvodina (in Serbian). *Radovi – Geoinstitut*, Belgrade.

Nikic, Z. & Vidovic, M. (2007). Hydrogeological Conditions and Quality of Ground Waters in Northern Banat, Pannonian Basin. *Environmental Geology*, Vol.52, No.6, pp. 1075-1084, DOI: 10.1007/s00254-006-0547-z.

Pravilnik o Higijenskoj ispravnosti vode za pice (Regulations in Drinking Water – in Serbian). Official Gazette SRY No.42/98 and 44/99.

Rakic, M. (1978). Textual Explanation of the Map Sheet Bela Crkva at Scale 1:100,000 (in Serbian), Geological Institute, Belgrade.

Slimak, T. (2005). Report on the Pavlis Water Resource (in Serbian). Documentation Fund, Institute Institute for Water Management „Jaroslav Cerni".

Spasic-Jokic, V. & Gordanic, V. (2010a). Geochemical Radiation Load in the Vrsac Hills, *Proceedings of the 15th Congress of Geologists of Serbia*, pp. 651-655, Belgrade, Serbia.

Spasic-Jokic, V. & Gordanic, V. (2010b). Monte Carlo Calculation of Ambient Dose Equivalent and Effective Dose from Natural Radionuclides in the Soil of Vojvodina Province of Serbia. *Third European IRPA Congress*, Helsinki.

UNSCEAR Report of the United Nationas Scientific Committee on the Effects of Atomic Radiation to the General Assembly 2000: United Nations, New York.

Vidovic, M. (2007). Water Quality, In: *Report, Geochemical-Geoecological Atlas at Scale 1:50,000 of Radioactive and Other Elements in Vrsac-Bela Crkva Area* (in Serbian), Gordanic, V. & Jovanovic, D. , 2007, Geological Institute of Serbia, Belgrade.

Vukovic, A. (1965). Textual Explanation of the Map Sheet Vrsac L34-103 (in Serbian), Institute for Geological and Geophysical Explorations, Belgrade.

Vukovic, M. (1987). Study and Prediction of Safe Water Supply to Vrsac and Nearby Communities to 2005 (in Serbian). Technical Documentation Fund, Institute Institute for Water Management „Jaroslav Cerni".

Zupunski, Lj., Spasic-Jokic, V., Trobok, M. & Gordanic, V. (2010). Cancer Risk Assessment After Exposure from Natural Radionuclides in Soil Using Monte Carlo Techniques. *Environmental Science and Pollution Research*, Vol.17, No.9, pp. 1574-1580. DOI: 10.1007/s11356-010-0344-9.

Conceptual Models in Hydrogeology, Methodology and Results

Teresita Betancur V., Carlos Alberto Palacio T.
and John Fernando Escobar M.
University of Antioquia
Colombia

1. Introduction

The foundation of model analysis is the conceptual model. A conceptual model in hydrogeology is a representation of the hydrogeological units and the flow system of groundwater.

Hydrogeology is far from being a typical quantitative science. Its models and predictions are only hypotheses; they rarely can be proven. There is rarely any proof of hydrogeologic hypotheses. Each model and each prediction is a hypothesis and there are rarely true tests of these predictions. Hydrogeology is mostly a descriptive science that attempts to be as quantitative as possible regarding descriptions, but without the possibility (in many or most cases) of guaranteeing the accuracy of predictions. Indeed, hydrogeologists should strive to make models more quantitative (rather than qualitative) in order to answer real management questions more precisely. But after answering such a question, what proof is there that their particular model is correct? There may be none (Voss, 2005).

A conceptual model is necessary in order to obtain a numeric model. Many aspects of the conceptual model are not possible to represent in numerical model, because the hydrogeological systems are very complex (Wagener, et al. 2007). A review about hydrologic models is in Gosain (2009). Sivakumar (2007), considers three main aspects in the models: processes, scale and objectives.

Groundwater exploration is defined as "all operations or work that allow the localization of aquifers or underground reservoirs from which water can be obtained in suitable quantity and quality for the intended purpose" (Custodio and Llamas, 1997.) The results of hydrogeological explorations are translated into conceptual models. As shown by Andersson and Woessner (1992), a conceptual model is a pictorial representation of the flow system of groundwater, often in the form of a block diagram or cross section. Conceptual models also include the characteristics of the hydraulic parameters of each unit, the positions of the phreatic and piezometric surfaces and also groundwater flow conditions. Besides these things, recharge areas and processes must be identified and reserves must be evaluated. The purpose of creating a conceptual model is to simplify the issue being examined and organize the data so the system can be analyzed effectively. Simplification is necessary because a complete reconstruction of the system is impossible. A conceptual model gives the basic idea or constructed understanding of how systems and processes operate (Bredehoeft, 2005).

As with all models, the quality of hydrogeologic models depends on the quality of the information that can be gathered for its construction, which in turn depends on the availability of financial resources.

It is important to note that a hydrogeological model contains many qualitative and subjective interpretations and proof of its validity can only be achieved by implementing specific research techniques and then constructing a numerical model and comparing the results from the simulation with observations from the field.

Bredehoeft (2005), referring to the certainty of conceptual models, uses the term SURPRISE and relates it to the situation in which the collection of new data invalidates an original conceptual model. Surprise may come from revision of the scientific theory or as a result of new information obtained on a particular site. According to Bredehoeft, surprise occurs in 20-30% of cases studied, indicating that it is not easy to build an appropriate hydrogeological

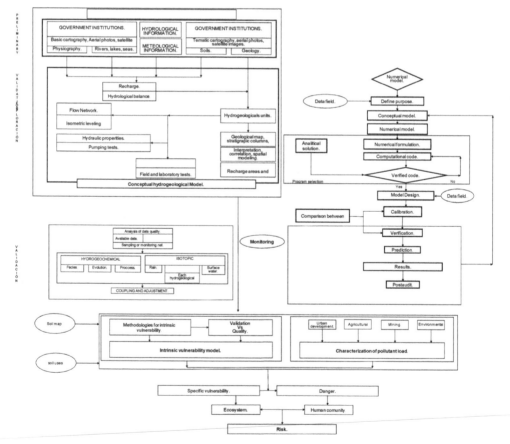

Fig. 1. Hydrogeological models Methodology

model. Carrerea et al. (2005) assert that given the uncertainty inherent in knowledge of the nature of subterranean media for those who deal with information which is quantitative (hard) but also qualitative (soft), it is possible to have several conceptual models for the same system.

The nature of hydrogeological variables is such that they can be interpreted with Digital Terrain Models (DTMs) obtained through geostatistical techniques. Geostatistical analysis includes: a description of information that is intended to quantify the relationship between measures of the same attribute in two locations, the modeling of spatial continuity represented by the variogram $(\gamma(h))$ – the cornerstone of geostatistical modeling, spatial prediction and assessment of uncertainty.

The coupling of hydrogeological research and exploration (Figure 1) involves applying a set of methodologies in search of a conceptual model. This methodological ensemble includes: basic exploration, numerical modeling, hydrochemical characterization, isotopic analysis, etc. Monitoring is the mechanism by which new geodata are obtained, and geographic information systems provide functional tools through spatial analysis to process and synthesize information (Betancur, 2008).

2. Conceptual models in hydrogeology, methods and techniques

A conceptual model is a reality representation, a conceptual model is a hypothesis; in order to obtain a good model in hydrogeology is necessary to join exploration techniques with validation methods. Then the tools for the management are necessary.

In this apart, five topics are considered: i) Hydrogeological data and information, ii) Numerical modeling, iii) Hydrogeochemical characterization and Isotope Hydrology, iv) Risk evaluation, and v) Geodata and geomodeling

2.1 Hydrogeological data and information

A hydrogeological model shows the surface distribution, geometry and hydraulic properties of the aquifer and the sources, zones and magnitude of the recharge as well as the quality of the underground water. The model requires for its construction prior information and knowledge of the physiographic, hydrographic, climatic, soils and geologic characteristics of the region (Figure 2). Available data and the information that can be extracted from them are the input elements to carry out an analysis procedure that can produce results to obtain the desired model.

To carry out local hydrogeological studies, a characterization of the physical medium must be made. Slope maps and Digital Terrain Models (DTM's) and three-dimensional views that represent the topographic surface can be obtained using the spatial analysis capabilities of GIS software. Photo interpretation, image processing and control field are all elements whose interpretation, in conjunction with review of papers and books make possible to obtain adequate physiographic and geological characterizations, and characterization of the soil in the study area. Descriptive statistics applied to the hydrometeorological information in the context of regional weather dynamics provide criteria for determining climatic conditions.

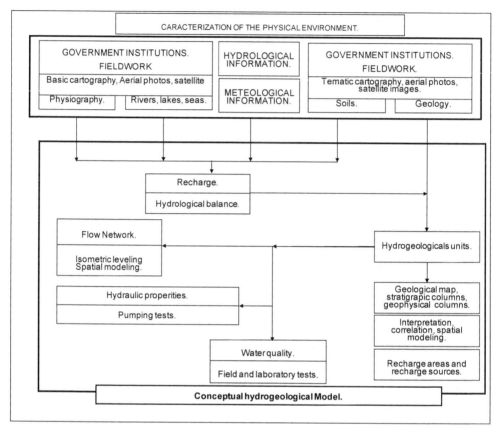

Fig. 2. Hydrogeological exploration methodology

Identification of the hydrogeological units is based on analysis of geology, stratigraphic information taken from inventories of water sampling points, and on geoelectrical data and its later correlation considering correspondence in character and position. For this, textural features, grain size, thickness, continuity and relative position in depth are taken into account for the stratigraphic units of each column.

In respect to the definition of the geometry of the aquifer units and their hydraulic properties and flow, geostatistics is recomended as the better method (Mejia et al., 2007). It is assumed that geostatistical analysis is a multistep process that requires the use of specialized software and includes;

i.   exploratory data analysis beginning with the use of descriptive statistics to determine basic facts;
ii.   evaluation of the norm and possible changes needed to achieve it;
iii. identification of trends and anisotropies;
iv.   construction of a semivariogram and computation of range, plateau and value of the nugget effect between other elements;

v.   adjustment of theoretical models to said variograms;
vi.   analysis of the nugget effect;
vii. evaluation of grouping (clusters) and
viii. cross-validation until finally managing to create the desired surface.

In light of these conceptual considerations, the process can be implemented using the Geoestatistical analyst module available through software from ArcGIS version 9.1. Kriging is a procedures very used in order to model geological characteristics. Kriging have not a level of precision that would permit that, for each known point, the predicted value be equal to the observed; however surfaces represented with Kriging have clear physical and stratigraphic senses and in consequence a clear hydrogeological sense. For this reason, results obtained by geostatistics are preferred over ones provided by interpolation models such as Spline or IDW.

To determine hydraulic properties, the use of artificial tracers is a good alternative for establishing conductivity and transmissivity conditions; chlorine, fluorescence and bacteria are the most commonly used substances (Divine and McDonnell, 2005). However, this technique has a little use in regional studies. The most common method to obtain hydraulic properties is the pumping tests; which require considerable funds, at least three observation wells and adequate space.

The most critical aspect in the implementation of the hydrogeological exploration projects is the assessment of recharge. In a general context, there are various methods for estimating the amount of water entering an aquifer: direct methods describe recharge as a mechanism of water percolation from the soil to the aquifer, while indirect methods are those that use variables that describe and represent the flow of water through the soil. In direct procedures, measurement devices such as lysimeters or environmental or artificial tracers are used, while with indirect procedures, it is necessary to determine the relationship between flow and the recharge. Of indirect procedures, water balance, which uses precipitation as the main input variable to the system, is the most commonly used.

The water balance equation is a ratio of conservation of mass in which the system inputs are equal to output plus the change in storage. In the long run, it is assumed that the change in storage is not very significant, but at scales of times of a month or more its magnitude is considerable (Equation 1).

$$P = Q + ETR + \Delta R \qquad (1)$$

Where:
P: Precipitation (mm).
Q: Direct Runoff (mm).
ETR: Actual evapotranspiration (mm).
Δ R: Change in storage (mm).

Total storage consists of underground storage and of water storage in the soil, which is available for use by vegetation. There are more complex alternatives for representation that characterize different layers of soil, in each of which exist different processes of hydraulic transport, and various energy transfers.

2.2 Numerical modeling

Besides beings a simulation tools, as they have normally been considered, in hydrogeology, numerical models offer a way to advance the understanding of groundwater systems. Models provide a framework to systematize information from the field and to answer questions about the functioning of an aquifer. They can call the attention of those who make the model to the occurrence of phenomena that had not been considered before, and can help identify areas where additional information is required. Numerical models can be exploratory in nature, and as tools for exploration, their construction may go along with the task of building a conceptual model from the moment information collection begins, through the process of investigation and can continue whenever new information is received or new analyses are applied for the validation of a hydrogeologic system.

Protocols for numerical modeling are clearly established. Up until the time of calibration, modeling is a method of continuous confrontation between available hydrogeological information or new information that is collected and the results that are obtained. The almost simultaneous development of a conceptual and numerical model decreases the level of uncertainty of them both, the interaction between them making them more solid in the sense that a higher level of reliability is acquired through the ongoing back-and-forth of information. This allows for better planning of new goals in the advancement of knowledge. Later, the numerical model becomes the tool for simulation and prognostication of possible scenarios.

From the system of partial differential equations governing the flow of groundwater, analytical solutions can only be applied to very simple and homogeneous systems. In order to solve that situation, a system of nodes is imposed for analysis, and according to the conditions of numeric modeling. A numeric model is obtained for a case of interest that represents the aquiferous medium being considered. The numerical methods that can be used to model a system include: finite differences, finite elements, finite volumes, integrated finite differences, boundary element methods and analytical elements, among others. Of all these, finite differences, finite elements and finite volumes are the procedures currently used most often to solve flow problems.

The mathematical combination of the equations of mass balance and Darcy's Law lead to equation 2, which describes the flow of groundwater.

$$\frac{\partial}{\partial x}.(K_x.\frac{\partial h}{\partial x})+\frac{\partial}{\partial y}.(K_y.\frac{\partial h}{\partial y})+\frac{\partial}{\partial z}.(K_z.\frac{\partial h}{\partial z})-W=S_s.\frac{\partial h}{\partial t} \qquad (2)$$

Where K_x, K_y, K_z are the hydraulic conductivity of the medium in the x,y,z, $\delta h/\delta x$, $\delta h/\delta y$, $\delta h/\delta z$ is the hydraulic gradient; S_s is the coefficient of specific storage; $\delta h/\delta t$ is the variation in piezometric charge over time; and W the ingresses and egresses of water caused by effects that are external to the aquifer system.

When the groundwater flow does not vary over time, $-\delta h/\delta t = 0$ - states that the system is in a steady or permanent state, otherwise it is considered to be a transient state. Flow simulation in a transient state assumes temporal discretization of hydraulic heads over intervals of time during which boundaries remain unchanged. These intervals of time intervals of simulation, will be discussed further on, and are known as periods of stress.

Equation 2, along with a number of conditions for piezometric heads or flow at borders and initial conditions, constitutes a mathematical representation of the conditions of groundwater flow. Solving the equation provides the values for the heads as a function of space and time, $h(x,y,z,t)$. Detailed discussions of numerical modeling of groundwater flow can be found in the writings of Wang and Anderson (1982) and Anderson and Woessner (1992).

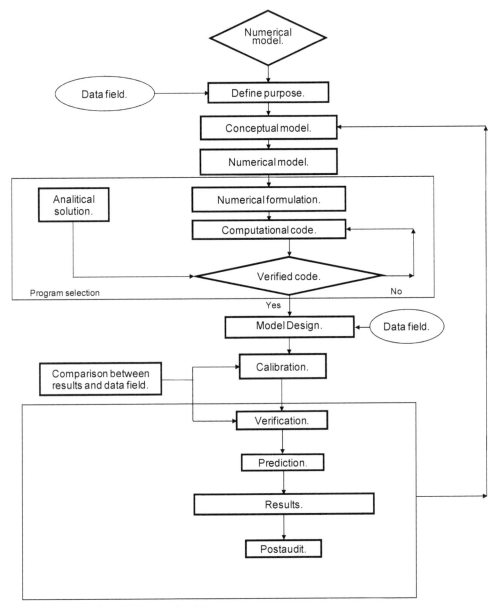

Fig. 3. Numerical modeling methodology

When it has been established that numerical modeling is an appropriate tool for exploring or simulating a hydrogeological system, the next steps are the design and implementation of this model. The suggested steps in modeling protocol are presented in Figure 3. Hydrogeological units and their boundaries should be defined in the conceptual model. The computer software selected for use in the modeling should contain an algorithm to solve numerically the mathematical model; afterwards, the screen is designed and increments of time, boundaries, and starting conditions are selected. The purpose of calibration is to manage, by means of adjustments to parameter values and ranges of simulation, to have the model reproduce field conditions for flow and piezometric heads. Uncertainty in the calibration, boundary conditions and timing of the simulation is quantified by a sensitivity analysis. In general, every modeling exercise should reach this stage. Only rarely is there sufficient information to verify a model, and long-term validation and monitoring— postaudit, is not necessarily a step in the protocol.

2.3 Hydrogeochemical characterization and Isotope Hydrology

In the field of hydrogeology, differences in the chemical and isotopic composition of water are used to check infiltration into groundwater, recognize leakage between aquifers, define areas of saltwater intrusion, assess baseflow contributions to surface currents and investigate recharge conditions through the unsaturated zone. The chemical characteristics of water allow identification of the origin and movement of solutes through subterranean systems. Environmental tracers help at the investigators understand the phenomenon of recharge, establish flow conditions and time scales in relation to groundwater permanence in the aquifer, and have enormous potential for helping to assess sustainability and vulnerability. Information from tracer isotopes and elements in solid and mineral phases has also helped investigators to understand paleohidrological conditions.

In its early stages, hydrogeological exploration provides information on water-rock interaction and thus enables the formulation of hypotheses to lead to validation of the conceptual model and understanding of the evolution of the aquifer. Isotopic characterization of a hydrogeological system is based on earlier hydrogeochemical

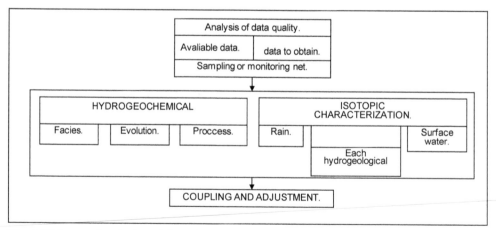

Fig. 4. Hidrogeochemical model methodology

characterization, and the two types of characterization are coupled in a cycle of confrontation and formulation of hypotheses that ultimately results in a validated conceptual model or a new conceptual model which must be subjected to new tests of validity. Questions are formed from hydrochemistry which are hoped to be answered by the isotopic data. For its part, the isotopic information may lead to new interpretations of available hydrogeochemical information.

The design and implementation of a network of sampling or monitoring for analysis of chemical and isotopic characteristics of rainwater, surface water and groundwater is based on a preconceived hydrogeological conceptual model. Interpretation of the data allows for an adequate characterization of the system, and from this, adjustment and validation of the conceptual model (Figure 4).

2.4 Risk pollution evaluation

The information contained in the hydrogeological model allows evaluation of the intrinsic vulnerability using different procedures, and knowledge of groundwater quality conditions enables the validation of methods to establish what method will be the most appropriate in a particular region. Knowledge of land use and the main economic activities carried out by the population provide the basis for assessing the potential or actual pollution load that could threaten the quality of water stored underground. The danger of contamination of groundwater is deduced from the relationship between intrinsic vulnerability and pollution load. Once the danger of contamination has been assessed, it is necessary to continue to a calculation of risk, for which an analysis of the susceptibility of the population to be adversely affected by a groundwater contamination must be undertaken (Figure 5).

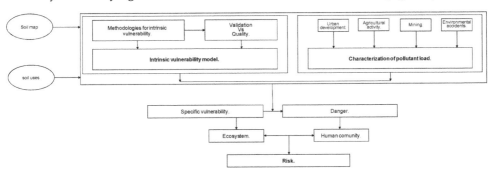

Fig. 5. Risk Evaluation methodology

2.5 Geodata and geomodeling

The establishment of geo-informatics as a technique for the analysis of information has led to the emergence of new ways of seeing and understanding elements and processes that occur in nature. These new ways of seeing and understanding are based on the possibility researchers have of sharing data and methods in order to represent, analyze, understand and model phenomena that vary over time and space, making use of explicit spatial objects and the options for data storage, integration, processing and analysis that are offered by Geographic Information Systems (GIS) and other tools such as remote sensing.

3. Conceptual models in hydrogeology: A case study

After seven years of study in the region of Bajo Cauca of Antioquia, Colombia, development and implementation of this methodology has allowed the construction and validation of a conceptual hydrogeological model and implementation of a monitoring network that continues to operate, providing new data and enabling new interpretations. Current knowledge of this hydrogeological system has offered great benefits to the process of decision making in the management of water resources, now featuring a risk assessment model for contamination of groundwater and moving towards formulating an Environmental Management Plan for the Aquifer (Betancur, 2005)

The Bajo Cauca of Antioquia is located between the last spurs and the foothills of the Western and Central ridges of the Colombian Andes mountain chain system. The alluvial plain of this region has an area of 3,750 km² and it is crossed by the Cauca, Man, Nechí, and Cacerí rivers (Figure 6).

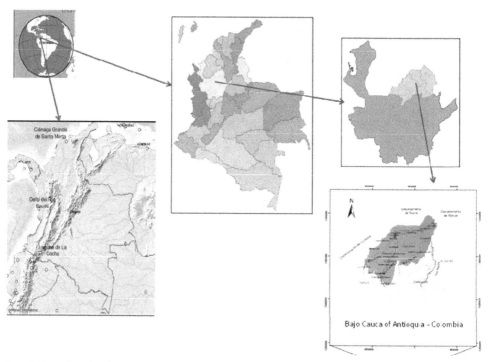

Fig. 6. Bajo Cauca of Antioquia, Study Area

The Bajo Cauca of Antioquia aquifer system is an important system. First, it is vital for communities as a water supply source; second, there is an interrelation between surface water and groundwater. Third, many wetlands depend on this resource. This aquifer system consists of three hydrogeological units: a free aquifer, an aquitard, and a confined aquifer.

3.1 Conceptual model

The aquifer system of the Bajo Cauca of Antioquia is comprised of three hydrogeological units (Figure 7); an unconfined aquifer known informally as hydrogeological unit U123, an aquitard, unit U4 and a confined aquifer, unit U5 (Betancur et al., 2009).

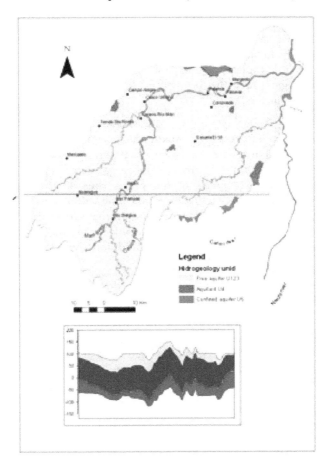

Fig. 7. Lower Cauca of Antioquia Study Area

The unconfined aquifer U123 consists of alluvial deposits from the rivers Cauca, Man, Nechi and Caceri and partially consolidated saprolite from Tertiary sedimentary rock of the Superior Member of the Cerrito Formation. U123 covers the entire study area. Its thickest points, between the Cauca and Man rivers, are greater than 90 meters, and its depth ranges between 40 and 90 meters. Parallel to the course of the Caceri River and towards the confluence of the rivers Cauca and Nechi, the unit has a depth of up to 60 meters. To the north and west, on the border with the department of Córdoba and in the south on the Andean slope, this unconfined aquifer is considerably narrower—even less than 10 meters.

The aquitard U4 is located below U123 and is made up of the Middle Member of the Cerrito Formation. Despite low conductivity, there are various catchments there from which water is extracted. From the center of the study area, following an axis in the direction SW-NE, the thickness of U4 decreases from 100 meters until the unit disappears in the north where U5 emerges or in the south where it intersects the Paleozoic basement. The depth of U4 reaches 160 meters in the center, about 20 meters in the north and less than 10 meters in the south.

The confined aquifer U5, formed by the Lower Member of the Cerrito Formation, is a regional confined aquifer in Bajo Cauca of Antioquia. Its thickness ranges from 10 meters to over 100 meters. This little explored and exploited unit could be an important reservoir of groundwater for the region. As is the case with its thickness, the depth of U5 is uncertain, but surely surpasses 260 meters.

Spatial distribution of hydrogeologic units, geomorphological attributes of the landscape, hydrography, the type of covering, hydraulic characteristics of the soil and hydrometeorological conditions are all factors that influence recharge of an aquifer system. For the study area, three sources of recharge were identified: 1) a recharge distributed across the whole expanse of the plains caused by direct infiltration of rainwater, 2) a recharge produced via the hydraulic interaction between principal bodies of surface water such as the Cauca and Man rivers and from some swamps and dams, and 3) an indirect lateral recharge provided by metamorphic rock of the system as much to the unconfined aquifer as to the confined aquifer. In addition, there is a vertical connection between the units U123, U4 and U5.

The water balances for an average hydrologic scenario, dry season in 1997 and wet season in 1990 show values of 1273 mm/year, 982 mm/year and 1729 mm /year respectively.

With changes in level not exceeding 5 meters between winter and summer, the phreatic level of the unconfined aquifer is located near the surface, and groundwater flow directions are marked by major watersheds. Between the Man river and Cauca river, and between Cauca river and Caceri river, groudwater flows from phreatic levels located between 90 and 140 meters towards the major surface currents which contribute base flow. Also, from the north on the border with the department of Córdoba, groundwater flow is directed towards the Cauca River. To the west, on the left slope of the Man River, the flow towards the channel can only be approximated, along with a possible flow in the opposite direction in some places (Figure 8).

From a series of data about hydraulic conductivity for the unconfined aquifer, was obtained value for this parameter between 1 and 2 m / day, with areas which reach 3 m/day. These results should be used with caution pending more reliable testing for quantification.

From data about groundwater quality taken from the unconfined aquifer (U123), it is observed that 21% of deposits do not meet the quality conditions of color, 19.7% do not meet the quality of turbidity, 22% fail on alkalinity, 19.7% of samples exceed the permissible value of total iron and 62% are outside the acceptable range of pH. For DQO and nitrite, 13% and 11% respectively of samples were found to be outside the acceptable range. Some samples accumulate 4 or 5 parameters with values outside the norm. In all cases in which coliform bacteria were tested for they were found an undesirable situation in water for human consumption.

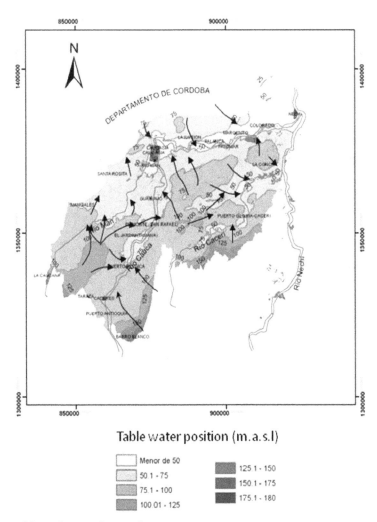

Fig. 8. Water table and groundwater flow

3.2 Numerical model

The scope in numerical modeling for the Bajo Cauca domains corresponds to the possibilities provided by available information. The level of confidence in the calibration for the unconfined aquifer is satisfactory, and although there were not sufficient piezometric monitoring data to allow for a thorough comparison with the transitional period, the results met expectations for the purposes to be achieved with an exploratory model (Figure 9). As for hydrogeological units U4 and U5, the numerical model lent support to ideas about the interactions that occur in the aquifer system and allowed for the identification of information necessary for the creation of a more robust conceptual model (Betancur et al. 2009).

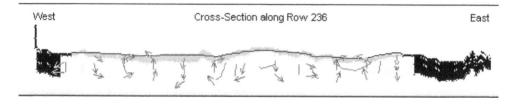

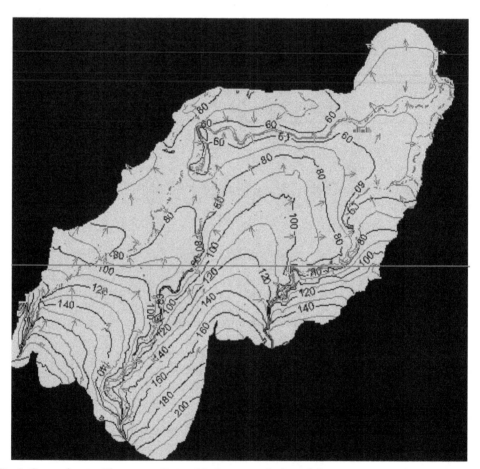

Fig. 9. Groundwater Flow according with the numerical model

3.3 A risk assessment for contamination of groundwater in the unconfined aquifer

The interaction between vulnerability and contaminating recharge allowed us to evaluate the degree of danger of contamination to which the unconfined aquifer of the Bajo Cauca of Antioquia is subject. Its relation to the presence of aquatic ecosystems and human communities allowed for a first assessment of the risks of being affected. The following characteristics were determined:

Vulnerability: from the application and validation of various techniques for assessing vulnerability, it was determined that the map obtained by DRASTIC is the most suitable for application in the case of the Bajo Cauca of Antioquia. According to results, the northwest has a low level of vulnerability and the rest of the system a medium level.

Contaminating recharge: urban-development activities— mainly sanitation systems without sewers and inadequate solid-waste disposal, as well as agricultural activities and mining in all cases generated a high Pollution Load Index (PLI) for the aquifer, an alarming situation If it is considered that a large number of sites where these conditions were found cause the source to be considered diffuse.

Risk of groundwater contamination: from the interaction between vulnerability and contaminating recharge it was established that urban development and agricultural activities generate moderate to high levels of danger, while for the regions of the Caceri and Nechí rivers, the mining industry reaches categories of extreme danger.

Risk: The Man and Caceri rivers, the latter in its upper and middle parts, are important habitats for fish and other species. Also, the complex system of wetlands in the Bajo Cauca, more than 150 wetlands, has been declared a protected area under the terms established within the national wetland policy, generated from Colombia's participation in the convention RAMSAR. It is well known that base flows are maintained by the flow of groundwater during droughts. Therefore, if contamination of subterranean waters of the unconfined aquifer is thought of as something that could impact aquatic ecosystems of the region, the danger is obvious. Not only are there zones with moderate, high and extreme levels of danger and low to moderate vulnerability levels, but it is clear from the direction of flow that streams and lentic bodies are receiving the groundwater. The risk to wetlands is imminent, and it is certain that the risk is not low.

No fewer than 150,000 people rely on the groundwater stored in the Bajo Cauca of Antioquia unconfined aquifer to satisfy their domestic needs. For this reason, the danger described for the aquifer is a danger for more than just the source of supply. So far it can say that the entire population that depends on the unconfined aquifer has a high probability of being negatively impacted with effects on their health, when the added risks of contamination of this hydrogeological unit from activities of urban development, agricultural activities and mining are considered (Gaviria, 2010).

3.4 Validation of the model using hydrogeochemistry and isotoes data

Facies of sodium bicarbonate and calcium bicarbonate to mixed bicarbonate were recorded in the waters of the unconfined aquifer. There are clear indications that the sodium bicarbonate facie samples were taken from wells with depths exceeding 50 meters and respond to the Upper Member of the Cerrito Formation, while samples with less evolved facies correspond to shallower waters with depths less than 20 m or that correspond only to alluvial deposits. It seems then that the unconfined aquifer has hydrogeochemical stratification which can be explained by assuming that water recently recharged by infiltration, which would have less residence time being exposed to a dynamic of rapid flow, is stored at higher levels, due not only to the recharge but also to the intense exploitation subterranean water undergoes at that level. The greater the depth

Trend LML con SPR-06, SPR-07 y SPR-08

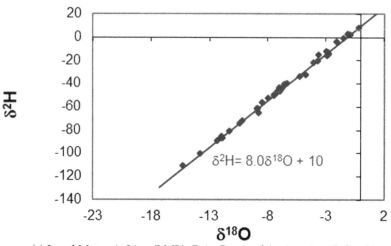

(a) Local Meteoric Line (LML), Bajo Cauca of Antioquia – Colombia.

Acuífero libre

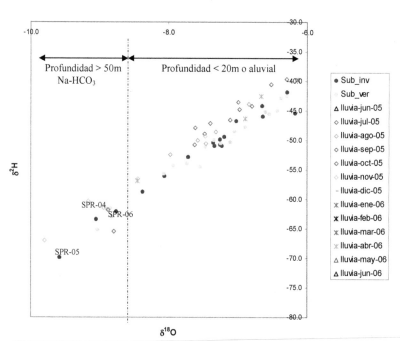

(b) Rain between 300 and 100 m.a.s.l and groundwater of the free aquifer

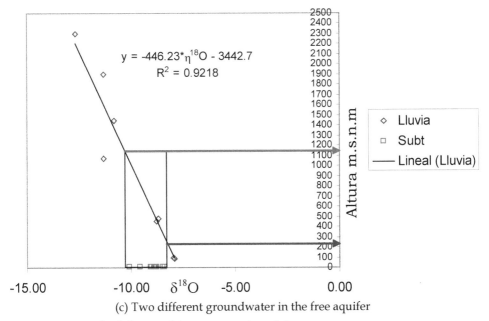

The equation shown in the figure:

$$y = -446.23 * \eta^{18}O - 3442.7$$
$$R^2 = 0.9218$$

(c) Two different groundwater in the free aquifer

Fig. 10. Isotope and validation of the conceptual model.

at which water is located, the more residence time it would have during which ion exchange processes could take place.

Very few water sampling points correspond to the confined aquifer, and most do so in regions where the synclinal structure of the system cause the confinement of the unit to be minimal. Only towards the center of the study area was a sufficiently confined sample taken with sodium bicarbonate composition.

From monthly rainfall samples taken for a year at eight precipitation stations a Local Meteoric Line (LML) very close to the Global Meteoric Line (GML) was obtained (Figure 10).

Individual analysis by season shows the effects of the seasons, altitude and continentality in the area. Besides the calculated values of excess deuterium, there were found to be three different sources of precipitation. The ratio of height vs. composition δ 18O weighted by season indicates a decrease of -0.21% δ 18O per 100m.

According to the conceptual model, recharge to the unconfined aquifer comes from direct infiltration from the surface or lateral recharge from the surrounding rock system in the south. Also, exchanges should be produced by winning or losing currents with surface sources. To test these hypotheses, isotopic ratios at points that correspond to this hydrogeological unit are considered with information about rain taken from stations located in these zones and with samples from bodies of surface water. Tritium data reported for two points of water corresponding to the unconfined aquifer with compositions of 1.5 ± 0.5 TU (Tritum Units), the same as that of rain, effectively indicate that recharge is currently ongoing.

All points of groundwater within a range of δ 18O values between -10.08 and -6.2 are located above or below the meteoric line. This situation suggests a dispersed recharge produced by rain water on occasions after it has undergone light evaporation. The closeness of the points of surface water, both to the line of rain and the points of groundwater, indicates, in principle, that during summer, flow volume is contributed by base flow from the unconfined aquifer, producing in the process an amount of evaporation and that in winter flow volume comes from base flow as well as direct runoff.

Considering seasonal variation in precipitation, it appears that groundwater recharge occurs from water precipitated mainly in the months of July, August and September.

For some groundwater samples that are more isotopically impoverished, in correspondence to altitudinal gradient, recharge would effectively be contributing between 292 and 1050 meters of height.

Tritium data for hydrogeological units U4 and U5 with values of 0.0 ± 0.5 TU indicate in principle that recharge has been occurring for at least 60 years.

4. Conclusion

- A model is a hypothesis; in order to obtain a good model in hydrogeology is necessary to join exploration techniques with validation methods. According with it, five methodology activities were propose: i) Hydrogeological data and information, ii) Numerical modeling, iii) Hydrogeochemical characterization and Isotope Hydrology, iv) Risk evaluation, and v) Geodata and geomodeling.
- Information and data are used in order to obtain geometric model, piezometric surfaces and groundwater flow directions, hydraulic properties, recharge, and quality water conditions. Numerical models can be used in order to explore or in order to simulate aquifer characteristics. Hydrogeochemical characterization and Isotope Hydrology are used in order to validate a conceptual model. Risk evaluation the impact over the population by the use of low quality groundwater.
- The establishment of geo-informatics as a technique for the analysis of information has led to the emergence of new ways of seeing and understanding elements and processes that occur in hidrogeology.
- In the region of Bajo Cauca of Antioquia, Colombia, development and implementation of this methodology has allowed the construction and validation of a conceptual hydrogeological model and implementation of a monitoring network that continues to operate, providing new data and enabling new interpretations.
- The aquifer system of the Bajo Cauca of Antioquia is comprised of three hydrogeological units; an unconfined alluvial aquifer known informally as hydrogeological unit U123, an aquitard, unit U4 and a confined aquifer, unit U5. U5 could be an important reservoir of groundwater for the region.
- For the study area, there are three sources of recharge: 1) a recharge distributed across the whole expanse of the plains caused by direct infiltration of rainwater, 2) a recharge produced via the hydraulic interaction between principal bodies of surface water such as the Cauca and Man rivers and from some swamps and dams, and 3) an indirect lateral recharge provided by metamorphic rock of the system as much to the

unconfined aquifer as to the confined aquifer. In addition, there is a vertical connection between the units U123, U4 and U5.

• Conductivity value for the unconfined aquifer, is between 1 and 2 m / day, with areas which reach 3 m/day.

• From the interaction between vulnerability and contaminating recharge it was established that urban development and agricultural activities generate moderate to high levels of danger, while for the regions of the Caceri and Nechí rivers, the mining industry reaches categories of extreme danger.

• The conceptual model to the hydrogeological system of Bajo Cauca of Antioquia was probed using numerical model, hydrochemical and isotopes data.

5. References

Anderson, M. P. and Woessner, W. W., 1992. Applied Groundwater Modeling. Simulation of flow and advective transport. Academic Press, San Diego. 381 p.

Bredehoeft, J., 2005. The conceptualization model problem—surprise. Hydrogeology journal, 13:37–46

Betancur, T.; 2005, estado actual y perspectiva de la investigación hidrogeológica en el bajo Cauca antioqueño, Boletín de Ciencias de la Tierra No. 17, Universidad Nacional, Medellín, pp. 97-108

Betancur, T., 2008. Una aproximación al conocimiento de un sistema acuífero tropical. Caso de estudio: Bajo Cauca antioqueño. Tesis doctoral, Universidad de Antioquia, 221p.

Betancur, T., Palacio, C. 2009. La modelación numérica como herramienta para la exploración hidrogeológica y construcción de modelos conceptuales (caso de aplicación: bajo cauca antioqueño) . En: Colombia. Dyna - Medellin, Colombia ISSN: 0012-7353 v.76 fasc.N/A p.39 – 49.

Betancur, T., Mejia, O., Palacio, C. 2009. Modelo hidrogeológico conceptual del Bajo Cauca antioqueño: Un sistema acuifero tropical. Revista Facultad De Ingenieria Universidad De Antioquia ISSN: 0120-6230.v.48 fasc.N/A p.107.

Carrera, J., Alcolea, A., Medina, A, Hidalgo, J. y Slooten, L., 2005. Inverse problem in hydrogeology. Hydrogeology journal 13:206–222

Custodio y Llamas 1997. Hidrología subterránea. Omega editores, Barcelona. 584 p.

Divine, C. E. y McDonnell, J. J., 2005. The future of applied tracers in hydrogeology. Hydrogeology journal, 13:255–258

Gaviria, J. I., 2010. Desarrollo y aplicación de una metodología para evaluar el riesgo a la contaminación de las aguas en un acuífero libre, caso de estudio: cuenca baja del río Man, Bajo Cauca antioqueño. Trabajo de Investigación: Maestría en Ingeniería con énfasis en Ingeniería Ambiental. Universidad de Antioquia. 164 p.

Gomez, A., 2010. Generación de un modelo hidrológico conceptual a partir de información secundaria: aplicación a la cuenca del río Man (Bajo Cauca antioqueño). Tesis de maestria, Universidad de Antioquia, 221p.

Gosain, K.A., 2009. Hydrological Modelling-Literature Review. Disponible en internet en: http://web.iitd.ac.in/~akgosain/CLIMAWATER. [Citado en 2010]. Reporte No 1. Octubre.

Mejía, O., Betancur, T.; y Londoño, L. 2007   Aplicación de técnicas geoestadísticas en la hidrogeología del Bajo Cauca antioqueño. Medellín, Revista DYNA edición 152, Vol. 74, pp. 137-150.

Sivakumar, B. 2007. Dominant processes concept, model simplification and classification framework in catchment hydrology. En: Stochastic Environmental Research and Risk Assessment. Vol 22. Pg 737–748.

Voss, C. I., 2005. The future of hydrogeology. Hydrogeology journal, 13:1–6.

Wagener, T; Sivapalan, M; Troch, P; Woods, R. 2007. Catchment   Classification and Hydrologic Similarity. En: Geography Compas. Vol 1, No 4. Pg 901-931.

Wang, H.F. y Anderson, M.P, 1982. Introduction to groundwater modeling. W.H. Freeman and Company, San Francisco, 237 p.

Groundwater Management by Using Hydro-Geophysical Investigation: Case Study: An Area Located at North Abu Zabal City

Sultan Awad Sultan Araffa

National Research Institute of Astronomy and Geophysics, Helwan, Cairo

Egypt

1. Introduction

Geophysical methods are applied to investigate geological structures, to identify promising areas, and to locate ore bodies by clarifying the distributions of physical properties in the Earth. Geophysics is a major discipline of the Earth sciences and a subdiscipline of physics, is the study of the whole Earth by the quantitative observation of its physical properties. Geophysical survey data are used to analyze potential petroleum reservoirs and mineral deposits, to locate groundwater, to locate archaeological finds, to find the thicknesses of glaciers and soils, and for environmental remediation. The theories and techniques of geophysics are employed extensively in the planetary sciences in general. Geophysical methods provide information about the physical properties of the earth's subsurface. There are two general types of methods: Active, which measure the subsurface response to electromagnetic, electrical, and seismic energy; and passive, which measure the earth's ambient magnetic, electrical, and gravitational fields. Geophysical methods can also be subdivided into either surface or borehole methods. Surface geophysical methods are generally non-intrusive and can be employed quickly to collect subsurface data. Borehole geophysical methods require that wells or borings be drilled in order for geophysical tools to be lowered through them into the subsurface. This process allows for the measurement of in situ conditions of the subsurface.

All geophysical techniques are based on the detection of contrasts in different physical properties of materials. If contrasts do not exist, geophysical methods will not work. Reflection and refraction seismic methods contrast compressional or shear wave velocities of different materials. Electrical methods depend on the contrasts in electrical resistivities. Contrasts in the densities of different materials permit gravity surveys to be used in certain types of investigations. Contrasts in magnetic susceptibilities of materials permit magnetic surveying to be used in some investigations. Contrasts in the magnitude of the naturally existing electric current within the earth can be detected by self-potential (SP) surveys. Seismic refraction surveys are used to map the depth to bedrock and to provide information on the compressional and shear wave velocities of the various units overlying bedrock. Velocity information also can be used to calculate in place small-strain dynamic properties of these units. Electrical resistivity surveys are used to provide information on the depth to bedrock and information on the electrical properties of bedrock and the overlying units. Resistivity surveys have proven very useful in delineating areas of contamination within soils and rock and also in aquifer delineation. Gravity and magnetic surveys are not used to the extent of seismic and resistivity surveys in geotechnical investigations, but these surveys have been used to locate buried utilities. Self-potential surveys have been used to map leakage from dams and reservoirs. Geophysical surveys provide indirect information. The objective of these surveys is to determine characteristics of subsurface materials without seeing them directly. Each type of geophysical survey has capabilities and limitations and these must be understood and considered when designing a geophysical investigations program. The analysis and processing of the geophysical data includes the mapping of surface features, the application of mathematical filters, the drawing of the geophysical grids, image processing techniques, modeling and mapping of the geophysical anomalies, etc..

Regional variations in density, resistivity, susceptibility, radioelement composition or reflectance spectra measured by geophysics can all be used to map geology or geomorphology. Large areas can be flown at a relatively low cost compared to ground surveys. The multiple sets of geophysical data can be combined with known geology to create regional geology maps and to develop priorities for follow-up on the most highly prospective ground. Interpreted geological structures such as shear zones, unconformities, or contacts may be strong indicators of economic mineralization. Mapping geology with geophysics has the power to extend geological knowledge into areas where the geology is covered, by transported sediment, for example. If the geophysical interpretation is constrained by regional geology, or spot observations of local geology, an effective geological map can be created for the entire survey area. Combining geophysical methods enhances the accuracy of the map(s) generated which they are more than the sum of the parts. Many rocks may be similar in one geophysical parameter, but vastly different in another.

Numerous geologic structures that can focus ground water flow can be identified by various geophysical surveys. Aquitards can be identified by direct current or electromagnetic resistivity studies. Clay lenses are less resistive than sand and gravel bearing aquifers, unless contained pore water is ion rich or highly acidic, in which case the aquifers may be less resistive. Reflection seismic and ground penetrating radar may distinguish subtle velocity and resistivity changes related to distributions of aquifers and aquitards in sedimentary environments. Pore fluid with properties that vary laterally and with depth cause some uncertainties in aquitard delineation. Lithologic boundaries and faults may be associated with porosity contrasts that may concentrate ground water flow. Porosity contrasts are associated with density, resistivity, and acoustic velocity contrasts, and some may have associated gamma-ray contrasts. In most cases, increased porosity lowers density and resistivity, the latter resulting from an increase of conductive pore fluid. In crystalline bedrock, faults and lithologic boundaries commonly have density and magnetic contrasts related to rock composition but ambiguities may result from subtle mineralogical changes in associated rocks. Bedrock topography, including structural highs and buried channels, that influence ground water flow may contrast with overlying sedimentary and alluvial cover in having higher density, magnetization, and resistivity. These contrasts may be identified using gravity, resistivity, ground-penetrating radar, and seismic refraction and reflection. Anomalies may be small and ambiguities in defining bedrock topography or buried channels include equivalent amplitude anomalies from variable properties of both bedrock and overburden

Resistivity is a geophysical method that images the earth by measuring the potential generated by injecting DC electricity into ground. The resulting geoelectrical image show the distribution of the earth' resistivity, which can be related to different soil and rock type. Many authors discussed the theory and application of resistivity method such as Keller and Frischknecht (1996), Teloford et al (1990) and Ward (1990). Electrical resistivity imaging (ERI) has become an important engineering and environmental site investigation tool, Laurence and Mehran (2004). Resistivity images are created by inverting hundreds to thousands of individual resistivity measurements (e.g., Loke and Barker, 1996a,b) to produce an approximate model of the subsurface resistivity. Rijo (1984) illustrated a simple inversion algorithm that uses a data bank of forward solutions for a certain class of 3D models. Li and Oldenburg (1992) proposed an inversion method based on the Born approximation Sultan and Santos (2008), Santos and Sultan (2008) used 1-D and 3D resistivity inversion for groundwater exploration and engineering geology.

1.1 Electrical resistivity

The electrical resistivity of any material depends largely on its porosity and the salinity of the water in the pore spaces. Although the electrical resistivity of a material may not be diagnostic, certain materials have specific ranges of electrical resistivity. In all electrical resistivity surveying techniques, a known electrical current is passed through the ground between two (or more) electrodes. The potential (voltage) of the electrical field resulting from the application of the current is measured between two (or more) additional electrodes at various locations. Since the current is known, and the potential can be measured, an apparent resistivity can be calculated. The separation between the current electrodes depends on the type of surveying being performed and the required investigation depth. Electrical resistivity, also referred to as galvanic electrical methods, is occasionally useful for determining shallow and deep geologic and hydrogeologic conditions. By measuring the electrical resistance to a direct current applied at the surface, this geophysical method can be used to locate fracture zones, faults and other preferred groundwater/contaminant pathways; locate clay lenses, sand channels and locate perched water zones and depth to groundwater. A variety of electrode configurations or arrays e.g., Wenner, Schlumberger, dipole-dipole can be used depending on the application and the resolution desired. Typically, an electrical current is applied to the ground through a pair of electrodes. A second pair of electrodes is then used to measure the resulting voltage. The greater the distance between electrodes, the deeper the investigation. Because various subsurface materials have different, and generally understood, resistivity values, measurements at the surface can be used to determine the vertical and lateral variation of underlying materials.

Reducing electrical resistivity data is a simple process in which the apparent electrical resistivity is calculated by dividing the measured voltages by the applied current. The quotient is then multiplied by the geometric factor specific to the array used to collect the data. Once the apparent electrical resistivities have been calculated, the next step is to model the data in order geologic structure. The method used to model the apparent electrical resistivity data is specific to each data acquisition mode.

1.1.1 Sounding mode

The two most common arrays for electrical resistivity surveying in the sounding mode are the Schlumberger and Wenner arrays. Electrode geometry for both arrays is shown in Figure 1.

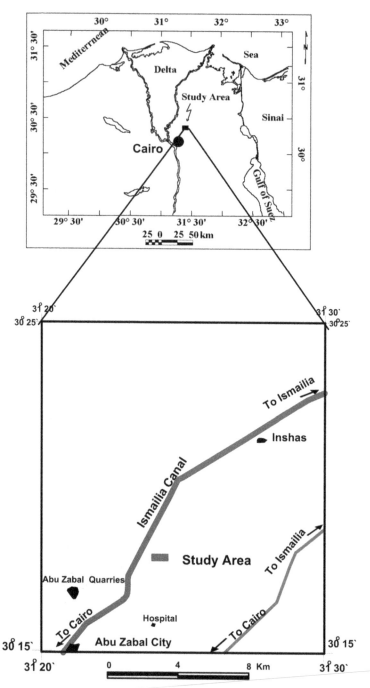

Fig. 1. Location map of the area.

Increasing the separation of the outer current electrodes, thereby driving the currents deeper into the subsurface increases the depth of exploration. Electrical resistivity sounding surveys measure vertical changes in the electrical properties of subsurface materials. The electrode spacing used for resistivity sounding is variable, with the center point of the electrode array remaining constant. Electrical resistivity data acquired in the sounding mode, using either the Wenner or Schlumberger array, can be modeled using master curves or computer modeling algorithms. When using master curves, the interpreter attempts to match overlapping segments of the apparent electrical resistivity versus electrode separation plots with a succession of two-layer master curves. This modeling method provides coarse estimates of the model parameters, is time consuming, and requires skill on the part of the interpreter. An alternative method of modeling sounding mode electrical resistivity data is to use readily available computer modeling software packages (Sandberg, 1990). There are a variety of different types of algorithms; some assume discrete electrical resistivity layers while others assume that electrical resistivity is a smooth function of depth. The discrete layer algorithms require interaction on the part of the interpreter, but allow for constraining model parameters to adequately reflect known geologic conditions. The continuous electrical resistivity algorithms are automatic, that is, they require no interaction on the part of the operator, and therefore geologic constraints cannot be incorporated into the models.

1.1.2 Profiling mode

The two most common arrays for induced polarization/electrical resistivity data collection in the profiling mode are the Wenner and dipole-dipole arrays. The electrode geometry for the Wenner array is the same as the sounding mode where the difference is that in profiling mode the entire array is moved laterally along the profile while maintaining the potential and current electrode separation distances. In the profiling mode, the distance between the potential and current dipoles (a dipole consists of a pair of matching electrodes) is maintained while the array is moved along the profile. Electrical resistivity profiling is used to detect lateral changes in the electrical properties of subsurface material, usually to a specified depth. Electrode spacing is held constant. Electrical resistivity profiling has been used to map sand and gravel deposits, map contamination plumes in hazardous waste studies, and used in fault studies.

In the dipole-dipole array (Fig.2), the typical field procedure is to transmit on a current dipole while measuring the voltages on up to seven of the adjacent potential dipoles. When the data collection is completed for the particular transmitter dipole, the entire array is moved by a distance equal to one dipole separation and the process is repeated. The most frequent source of inaccuracy in electrical resistivity surveying is the result of errors in the placement of electrodes when moving electrodes and/or expanding the electrode array. These distance measurement errors are easily detected on apparent electrical resistivity versus electrode separation curves and for this reason the apparent electrical resistivities should be plotted as the data is acquired in the field. A qualified field geophysicist will recognize these errors and direct the field crew to check the location of the electrodes. The second most common source of error in electrical resistivity surveying is caused by the electrical noise generated by power lines. The most effective means of reducing power line noise is to minimize the contact electrical resistance at the potential electrodes. This can be easily accomplished by using non-polarizing potential electrodes along with wetting the soil under the electrode with water. Non-polarizing electrodes are recommended instead of metal potential electrodes, because the metal electrodes generate electrical noise due to oxidation reactions occurring at the metal-soil (pore water) interface.

In the present study 2D dipole-dipole , 3D inversion for 1-D VES and ground magnetic survey were carried at the area located at the north of Abu Zabal city and at distance of 2km west Ismailia Canal (branch of River Nile) and cover an area of 0.5 km² to investigate stratigraghy , structures and groundwater occurrences (Fig.1).

2. Geology of the area

The surface geology was described by Geological Survey of Egypt (1998) as shown in Figure 2a where most the study area covered by Quaternary deposits which consist of alluvium, sand sheet and Nile silt and Nile mud. The Miocene deposits occupied the northeastern and southern parts of the area which represented by Hommath formation and composed of sandy limestone, sand, sandstone and clay (Fig.2a). Stratigraphically, the drilled boreholes reflect the rock types with depth in the area. The Quaternary deposits are represented Nile silt and mud, loose sands, sandstone, gravel with a thickness ranging from few meters to 20 m. The Miocene deposits are expressed limestone, sand, sandstone, clay and calcareous sandstone. The Oligocene deposits are represented by basaltic flow and sand and sandstone of Gabal Ahmar Formation. The depth of the upper surface of the basaltic flow ranging from 75 m at borehole no.7 to 128 m at borehole no.9. Gabal Ahmar Formation appeared at the borehole no.13 at depth of 113.5 m to the end of the borehole at depth of 202m (Fig.2b).

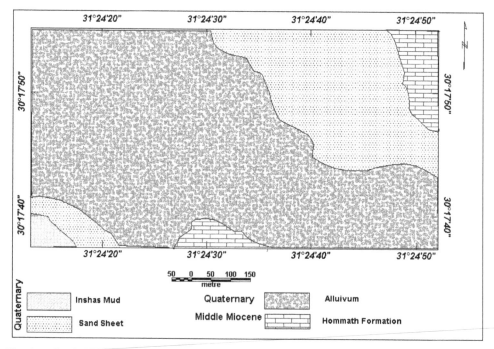

Fig. 2. a. Geological map of the study area

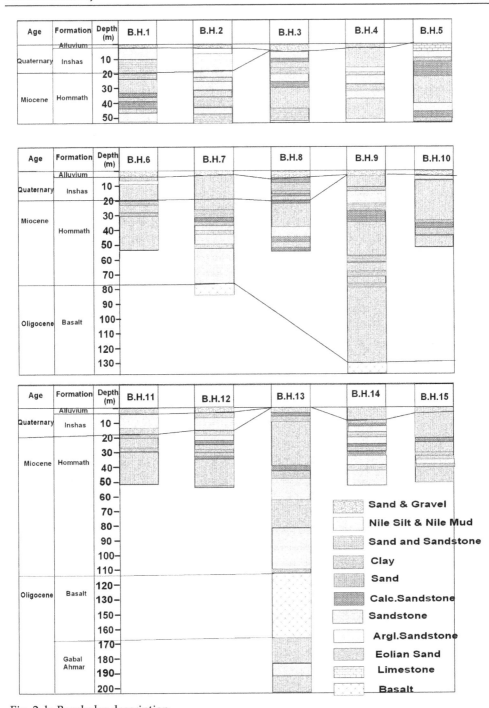

Fig. 2. b. Boreholes description.

3. Methodology

3.1 Shallow geoelectrical investigation tools

The shallow investigation tools in the present study were represented by measuring five dipole dipole sections. The length of each section is one kilometer and spacing between sections is 125 m (Fig.3). The spacing between current electrodes and potential electrodes is a multiple of the electrode spacing (a), in the present study (a) is equal 5 m. The depth of penetration is a function of the distance a (Edwards, 1977; Loke and Barker, 1996b). The dipole-dipole profiles were inverted using the RESINV2D software which produced an image of the electrical resistivity distribution in the subsurface based on a regularization algorithm (Loke and Barker, 1996b). The inverted dipole-dipole section along profile P1-P1` (Fig.4a) exhibits large variation in resistivities where the first half of the section divided into two parts, the first part is shallow depth ranging from 1.1 to 4 m and shows moderate resistivities ranging from 10 to 50 ohm.m corresponding to alluvium deposits, the second part is at depth ranging from 4 to 11.1 m and exhibits very low resistivities ranging from 1 to 10 ohm.m corresponding to Nile silt and Nile mud. The second half of the section reflects very high resistivities up to 750 omh.m corresponding to sand sheet and sandstone, the end part of the section reveals moderate and low resistivities corresponding to limestone and clay respectively. The second dipole-dipole section along profile P2-P2` (Fig.4b) reveals low resistivities at the first half of the section according to alluvium and Nile silt and Nile mud, the second half reveals high resistivities corresponding to sand and sandstone with some clay of low resistivities. The dipole-dipole section along profile P3-P3` (Fig.4c) indicates low resistivities for most section corresponding to alluvium and Nile silt and Nile mud, the last part of the section reveals high resistivities at depth ranging from 4 to 11.1 m corresponding

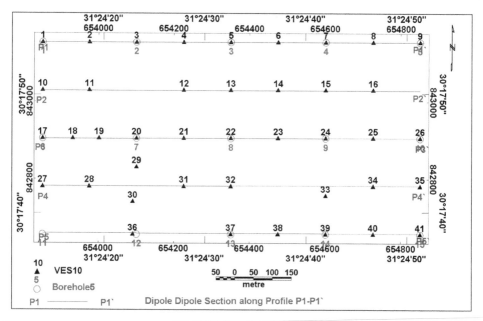

Fig. 3. Location map of geophysical measurements and boreholes

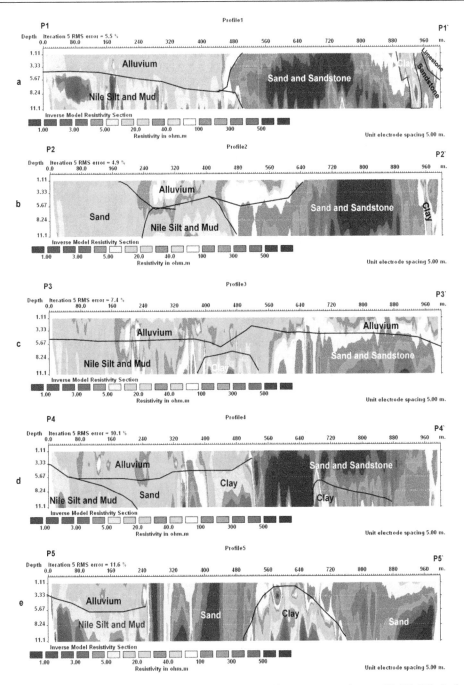

Fig. 4. Dipole-dipole section inverted using RES2DINV program; a. for profile P1-P1`, b. for profile P2-P2`, c. for profile P3-P3`, D. for profile P4-P4` and e. for profile P5-P5`.

to sand and sandstone. The dipole-dipole section along profile P4-P4` (Fig.4d) shows variation in resistivities ranging from low values according to clay, Nile silt and Nile mud to very high rsistivities of sandstone and sand. The last section along profile P5-P5` (Fig.4e) exhibits at the first and mid part from distance 500 to 720 m a very low resistivities ranging from 1-10 ohm.m according to clay, Nile silt and Nile mud, the mid part at distance from 250 to 500 m and the last part of the section reveals very high resistivities corresponding to sand and sandstone.

3.1.1 3-D representation for dipole-dipole data

The dipole-dipole data were represented by 3D slices at different depths of 1.1, 3.3, 5.7, 8.2 and 11.1 m (Fig.5). The slices at depth 1.1 and 3.3 m reveal low resistivities ranging from 5 to 100 ohm.m at the most of the area corresponding to alluvium deposits, the northeastern and southeastern parts are occupying by very high resistivity up to 750 ohm.m. The slice at depth of 5.7 m shows very low and low resistivities at the western half of the area corresponding to clay, Nile silt and Nile mud, the eastern half of the area is occupying by high and very high resistivities for sand and sandstone. The slices at depth 8.2 and 11.1m exhibit very low resistivities corresponding to Nile silt and Nile mud at the western part of the area, the eastern part is occupying by sand and sandstone of high resistivities.

3.2 Deep geoelectrical investigation tools

The deep geoelectrical tool is represented by measuring 41 vertical electrical soundings (VES) of AB/2 ranging from 5 m to 500 m using the Schlumberger configuration (Fig.3). The data were acquired using a SYSCAL–R2 resistivity meter.

3.2.1 3D resistivity inversion

The objective of the inversion process is to obtain a distribution of the electrical resistivity (model) whose response approaches the field data (apparent resistivity values) within the limits of data errors and that correlates well with all available data, especially the geological information (Santos and Sultan, 2008). The smoothness-constrained least-squares method has been widely used in 2-D and 3-D inversions of magnetotelluric, electromagnetic and geoelectrical data sets (DeGroot-Hedlin and Constable, 1990; Sasaki 1989, 1994 and 2001). The scheme adopted in this study is based on Sasaki (1994, 2001). The dc resistivity inverse problem can be expressed as

$$J \, \Delta p = \Delta d \tag{1}$$

where Δp is the vector containing the corrections to the model parameters p, $\Delta d = y^c - y^{ob}$ is the vector of the differences between the model responses and the measured data, and J is the derivative matrix (Jacobian) containing the derivatives of the model responses. In the present study forty-one were inverted by using a new approach, the full description for this approach was presented in Santos and Sultan, 2008. The results of the 3D inversion firstly represented by Figure 6 which shows the comparison between measured and calculated apparent resistivity curves. The misfit between data and model responses is a good fitting for most VES stations with rms error less than 10 % except the VES stations no.20, 39, 40 and 41 exhibit bad fitting.

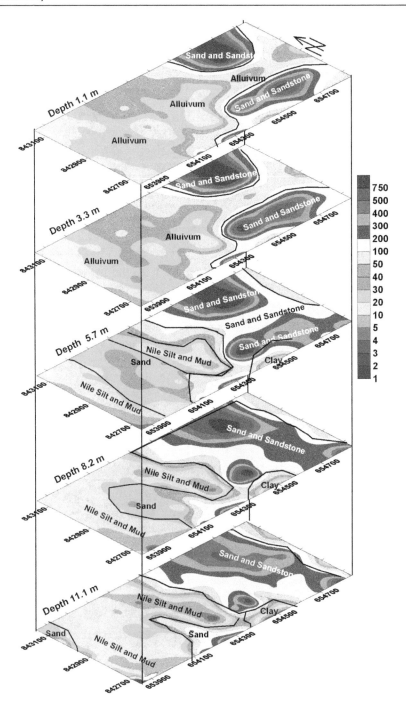

Fig. 5. 3D representation for dipole-dipole data

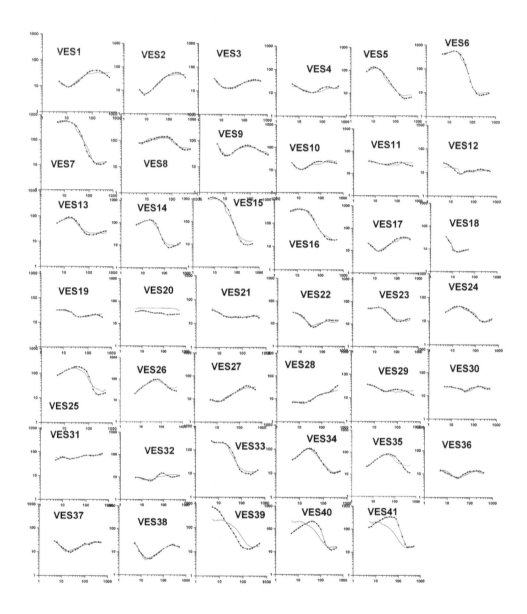

Fig. 6. Fitting of 3D VES inversion.

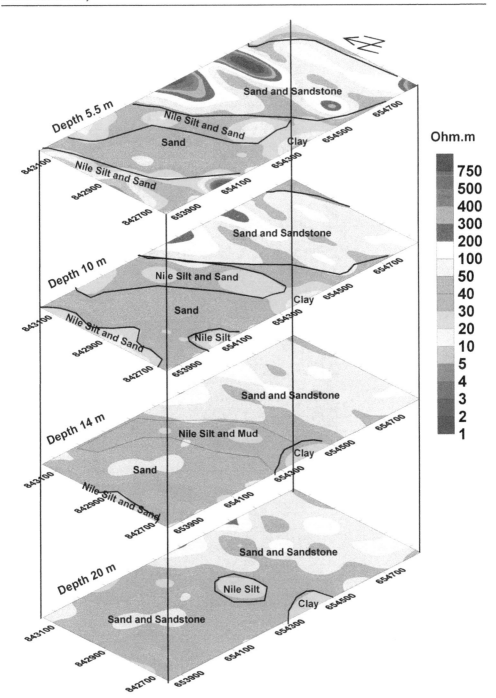

Fig. 7. 3D VES inversion slices at depth 5.5, 10, 14 and 20 m.

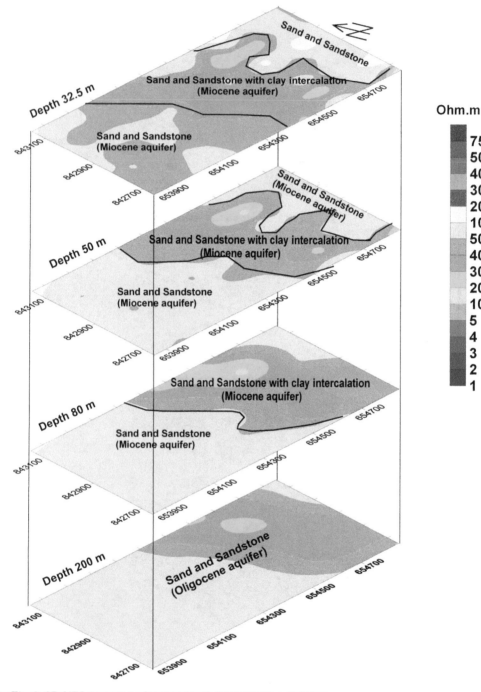

Fig. 8. 3D VES inversion slices at depth 32.5, 50, 80 and 200 m.

3.2.2 3D resistivity (VES) inversion slices

The resistivity depth slices for the results of 3D inversion for VES data indicate that the study area show large variation in resistivities according to lithologic composition. 3D slice at depth 5.5 m exhibits very low resistivities ranging from 4-20 ohm.m according to Nile silt and Nile mud at the western and central part of the area, the low resistivities ranging from 30-50 ohm.m occupying the eastern part corresponding to sand. The eastern part of the area reveals high and very high resistivities ranging from 100-750 ohm.m for sand and sandstone. The depth slice at 10 m exhibits low resistivities corresponding to sand, Nile silt and Nile mud at the western part of the area, the eastern part reflects high resistivities according to sand and sandstone. The depth slices at 14 and 20 m show variation in resistivities ranging from 20-200 ohm.m corresponding to sand, Nile silt, Nile Mud and sandstone. The Miocene groundwater aquifer appears at depth slice 32.5 at the most study area where the Miocene aquifer consists of sand and sandstone of clay intercalation at the central part and the western part, the aquifer composed of sand and sandstone of moderately resistivities. The eastern part represents the dry Miocene sand and sandstone of high resistivities ranging from 50-100 ohm.m. The slices at depths 50 and 80 m represent the Miocene aquifer which cover all the area and composed of clayey sand and sandstone at the central part but the eastern and western part occupying by sand and sandstone. The basaltic sheet is not represented in 3D VES slices where the depths of basalt ranging between 80 and 120 m and 3D slices were represented by slices at depths 5.5, 10, 14, 20, 32.5, 50, 80 and 200 m (Figs.7 and 8).

3.3 Hydrogeology of the study area

Hydrogeology is the branch of geology that deals with the occurrence, distribution, and effect of groundwater. Some hydrogeologcal studies were carried out on the study area such as flow measurements to get an idea of water flow along the Ismailia Canal course and quantity of water which seeps into the groundwater from the canal, the measurements were carried by Geological Survey of Egypt (EGSMA, 1998) along section of length 28 km between Abu Zabal south study area (upstream) to Belbies north study area (down-stream). The results of the measurement indicated that the velocity of water in Canal is 0.87 m/s, a section area = 51.5 m², amount of seepage = 45 m³/s for the upstream and an section area = 90 m², a velocity = 0.45 m/s, amount of seepage = 40.5 m³/s for the down stream. The measurements indicated also, the side flow input = 0 m³/s and output side flow = 4.13 m³/s. The measurements include also 10 tests for estimating the hydraulic conductivity, the results of tests indicated the hydraulic conductivity is 0.102, 0.265, 1.233, 1.427, 0.081, 0.061, 0.0815, 1.019, 0.112 and 0.122 cm/min. Water table was estimated through seven boreholes no. 1, 3, 5, 7, 9, 11 and 15 as shown in table (1).

The topographic map (Fig.9a) of the study area was constructed using elevation measurements which were carried out by using a Leica TC805 "Total Station" instrument of high resolution and good readability under all light conditions, where the study area is characterized by gentle slop from east (57 m) to west (35 m). Water table map (Fig.9b) shows two gentle gradients, the first from east to west and the second located at the southeast part of the area where the water table varied from 13.2 to 14.1 m. The water table map indicates that the recharge source is Ismailia canal from southeast part and Miocene aquifer at the eastern part. Figure 9c reflects the depth of water ranging from 21 at the northwest part to 49 m at the northeast part.

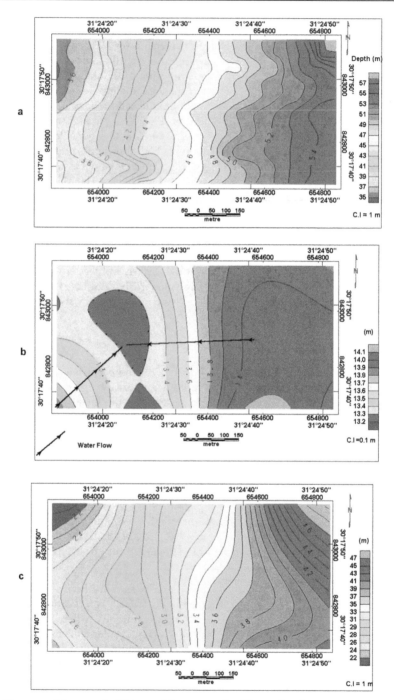

Fig. 9. a. Topographic map, b. Water table map and c. Water depth map

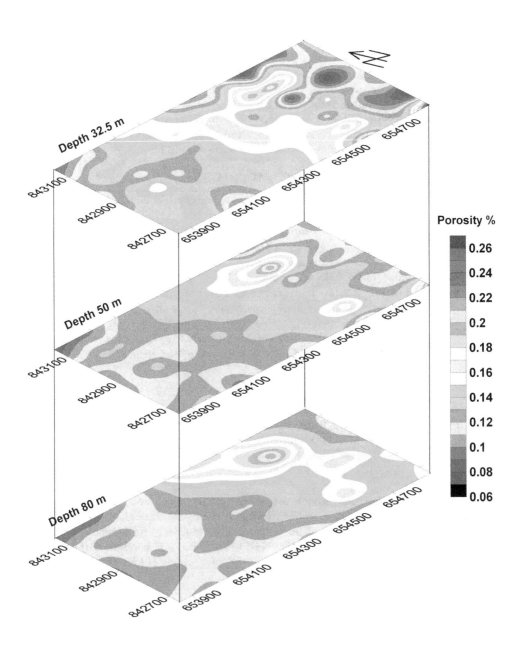

Fig. 10. Porosity percentage of the Miocene aquifer at depths 32.5,, 50 and 80 m.

3.3.1 Porosity estimation

The porosity of the aquifer for clay free can be estimated through the equation 2 (Mark and Uri, 2004)

$$\rho_w = \rho_b \phi^2 \qquad (2)$$

Where the ρ_b is the bulk resistivity of the rock, ρ_w is the resistivity of water within the pore space, ϕ is fractional porosity of the rock (approximately representing the volume of water filling the pore space), but the study area contains clay according to the boreholes results (Fig.2), Sen et al, 1988 summarized an equation for estimating the porosity through the bulk conductivity and groundwater conductivity for the aquifer contains clay (clayey sand and clayey sandstone), parameters in this equation are shown in equation 3

$$\sigma = \phi^m \{ \sigma_w + AQ_v/(1+CQ_v/\sigma_w)\} \qquad (3)$$

Where σ is the conductivity of the aquifer, σ_w is water conductivity, ϕ is porosity of the aquifer, m, A and C are constants Q_v the clay charge contribution per unit pore volume. Sen et al determined these parameters through 140 core samples as the following $m= 2$, $A=$ 1.93xm (mho/m)(1/mol), $Q_v = 2.04$ and $CQ_v = 0.7$ (mho/m). σ_w was estimated through boreholes which drilled in the study area, the average values of σ_w is 0.119 s/m , the results of porosity estimation of the Miocene aquifer were represented by Figure 10. The slice at depth 32.5 m (Fig.10a) reveals low porosity percentage at the eastern part ranging from 0.08 to 0.11 %, but the central part shows high porosity percentage ranging from 12 to 23 %. The porosity distribution at depth 50 m (Fig.10b) exhibits low porosity percentage at the eastern and western parts, the central part reveals high porosity 23 %. The porosity percentage at depth 80 m indicates that the most study area has high porosity percentage (13-23 %).

3.3.2 Hydrochemistry of the area

The hydrochemistry of the water has been described through six samples of water were taken from the boreholes and listed in table (1). The total dissolved salts (TDS) were represent by salinity map (Fig.11a) which indicates that the central part is occupying by high salinity (2200 mg/l) and the southwestern part reveals low salinity about 550 mg/l.

Water type

The type of water is detected through the representation of boreholes data through Piper trilinear diagram and ratios of main ions in water chemistry through Schoeller diagram, where theses diagrams indicate that the water type is Na-Cl-SO4 for boreholes 3, 7 and 9; Na-HCO3-SO4 for boreholes 11 and 15; Na-Cl-HCO3-SO4 for borehole 1 (Fig11b). The ratios of ions are higher for the borehole 9 and the lower ratios of ions are for Borehole 11 (Fig.11c). The total hardness have been estimated through the following equation (Reinhard, 2006)

$$\text{Hardness } (H_{tot}) = 0.14^*Ca^{+2} + 0.23^* Mg^{+2} \qquad (4)$$

The results of hardness estimation has been tabulated in table (1) , the result indicate that the total hardness of water is a slightly hard (17.1-60) , but the borehole 9 exhibits 108.5 mg/l (moderately hard).

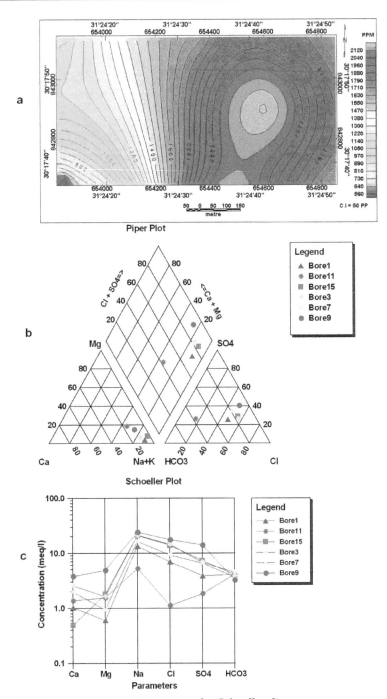

Fig. 11. a. Salinity map, b. Pieper Diagram and c. Schoeller diagram

	Borehole1	Borehole3	Borehole7	Borehole9	Borehole11	Borehole15
Cl⁻ mg/L	261	516.4	329.1	624.2	39.7	493
SO₄⁻ mg/L	184.8	326	314.1	677.3	88.6	350.3
HCO₃ mg/L	247.1	271.5	222.7	198.3	247.1	247.4
Na⁺ mg/L	307.9	488	375.2	553.2	121.4	482.2
K⁺ mg/L	9.1	13.8	8.9	15.7	7.8	9.9
Ca⁺⁺ mg/L	20.7	51.7	39.6	75.8	27.6	44.8
Mg⁺⁺ mg/L	7.3	19.8	11.4	59.3	18.7	21.8
TDS mg/L	1037.9	1687.2	1301	2213.8	550.9	1650.1
TH mg/L	330.5	348	290.3	593.5	351.2	314
PH mg/L	7.61	7.77	7.86	7.85	7.86	7.54
EC μ mho/cm	850	1250	1000	3250	700	1300
Total Hardness (mg/l)	27.551	50.194	46.596	108.461	16.705	54.056
Water table (m)	13.58	13.8	13.28	14.05	13.84	14.3
Total depth (m)	50	50	75	128	50	50
Elevation (m)	34	44.8	42	51.5	37.5	55.6

Table 1. Hydrochemistry data collected from drilled boreholes in the study area.

4. Conclusion

Through the interpretation of the geophysical and hydrogeologicat data we can concluded that the shallow part of the stratigraphic section of the study area consists of Nile silt, Nile mud and clay while the eastern part exhibits high resistivities corresponding to sand and sandstone for this reason the eastern part of the study area is suitable for constructions The depth of the Miocene aquifer is ranging from 32.5 to 80 m, the quality of the water is good and suitable for drinking and agricultural purposes where the salinity ranging from 550-2200 gm/1 and the total hardness is ranging between 16.7-108.5 gm/1. The depth of the basaltic sheet ranging from 75 to 250 m. The area dissected by fault elements of NW-SE, NE-SW and N-S trend.

5. References

DeGroot-Hedlin C. and Constable S.C., 1990. Occam's inversion to generate smooth, two-dimensional models from magnetotelluric data. *Geophysics*, 55, 1613-1624

Edwards, L.S., 1977. A modified pseudosection for resistivity and induced polarization. Geophysics, 42, 1020-1036.

Geological Survey of Egypt (EGSMA), 1998. Geology of Inshas Area, Geol.Surv. of Egypy, internal report

Geosoftw Program (Oasis Montaj), 1998. Geosoft mapping and application system, Inc, Suit 500, Richmond St. West Toronto , ON Canada N5SIV6

Keller, G.V., and Frischknecht, F.C., 1996: Electrical methods in geophysical prospecting, Pergamon Press.

Li, Y., and Oldenburg, D. W., 1992, Approximate inverse mappings in DC resistivity problems: Geo hys. J. Int., 109, 343-362.

Loke, M. H., and Barker, R. D., 1996a, Practical techniques for 3D resistivity surveys and data inversiton: Geophysical Prospecting, 44, 499–523.

Loke, M. H., and Barker, R. D., 1996b, Rapid least-squares inversion of apparent resistivity pseudo-sections using quasi-Newton method: Geophysical Prospecting, 44, 131–152

Laurence Bentley and Mehran Gharibi, 2004 : Case study, Two- and three-dimensional electrical resistivity imaging at a heterogeneous remediation site, Geophysics, Vol. 69, No. 3, P. 674–680,

Mark Goldman and Uri Kafri; 2004 : Hydrogeophysical Applications in Coastal Aquifers, Applied Hydrogeophysics, NATO Science Series, published by IOS Press, Amsterdam.

Monteiro Santos F.A and Sultan S.A.; 2008: On the 3-D inversion of Vertical Electrical Soundings: application to the South Ismailia - Cairo Desert Road area, Cairo, Egypt, Journal of Applied Geophysics, 65, pp 97-110

Reinhard Kirsch; 2004 : Groundwater Geophysics, A Tool for Hydrogeology, text book, ISBN 10 3-540-29383-3 Springer Berlin Heidelberg New York.

Rijo, L., 1984, Inversion of three-dimensional resistivity and induced-polarization data: 54th Ann. Internat. Mtg., Soc. Expl. Geophys., Expanded Abstracts, 113-117

Sasaki Y., 1989. Two-dimensional joint inversion of magnetotelluric and dipole-dipole resistivity data. Geophysics, 54, 254-262.

Sasaki Y., 1994. 3-D resistivity inversion using the finite element method. Geophysics, 59, 12, 1839-1848.

Sasaki Y., 2001. Full 3-D inversion of electromagnetic data on PC. Journal of Applied Geophysics, 46, 45-54.

Sen, P.N., Goode, P.A. and Sibbit, A., 1988. Electrical conduction in clay bearing sandstones at low and high salinities. J. Appl. Physics, 63 (10), 4832-4840

Sultan, S.A., Monteiro Santos F.A (2008) : 1-D and 3-D Resistivity Inversions for Geotechnical Investigation, J. Geophys. Eng. 5, pp 1–11

Sultan S.A , Santos F.A.M, and . Helaly A.S; 2011: Integrated geophysical analysis for the area located at the Eastern part of Ismailia Canal, Egypt, first online, Arabian geosciences journal, 4, 735-753

Telford, W.M., Geldart, L.P., and Sheriff, R.E., 1990: Applied Geophysics (2nd Edition), Cambridge University Press.

Thompson, D.T.T., 1982. Euler, a new technique for making computer assisted depth estimates from magnetic data, Geophs.V.47.pp.31-37.

Ward, S., 1990 : Resistivity and induced polarization methods, in Ward, S., Geotechnical and environmental geophysics, Vol.1: SEG Investigation in geophysics 6, 147-189

Permissions

The contributors of this book come from diverse backgrounds, making this book a truly international effort. This book will bring forth new frontiers with its revolutionizing research information and detailed analysis of the nascent developments around the world.

We would like to thank Gholam A. Kazemi, for lending his expertise to make the book truly unique. He has played a crucial role in the development of this book. Without his invaluable contribution this book wouldn't have been possible. He has made vital efforts to compile up to date information on the varied aspects of this subject to make this book a valuable addition to the collection of many professionals and students.

This book was conceptualized with the vision of imparting up-to-date information and advanced data in this field. To ensure the same, a matchless editorial board was set up. Every individual on the board went through rigorous rounds of assessment to prove their worth. After which they invested a large part of their time researching and compiling the most relevant data for our readers. Conferences and sessions were held from time to time between the editorial board and the contributing authors to present the data in the most comprehensible form. The editorial team has worked tirelessly to provide valuable and valid information to help people across the globe.

Every chapter published in this book has been scrutinized by our experts. Their significance has been extensively debated. The topics covered herein carry significant findings which will fuel the growth of the discipline. They may even be implemented as practical applications or may be referred to as a beginning point for another development. Chapters in this book were first published by InTech; hereby published with permission under the Creative Commons Attribution License or equivalent.

The editorial board has been involved in producing this book since its inception. They have spent rigorous hours researching and exploring the diverse topics which have resulted in the successful publishing of this book. They have passed on their knowledge of decades through this book. To expedite this challenging task, the publisher supported the team at every step. A small team of assistant editors was also appointed to further simplify the editing procedure and attain best results for the readers.

Our editorial team has been hand-picked from every corner of the world. Their multi-ethnicity adds dynamic inputs to the discussions which result in innovative outcomes. These outcomes are then further discussed with the researchers and contributors who give their valuable feedback and opinion regarding the same. The feedback is then collaborated with the researches and they are edited in a comprehensive manner to aid the understanding of the subject.

Apart from the editorial board, the designing team has also invested a significant amount of their time in understanding the subject and creating the most relevant covers. They scrutinized every image to scout for the most suitable representation of the subject and create an appropriate cover for the book.

The publishing team has been involved in this book since its early stages. They were actively engaged in every process, be it collecting the data, connecting with the contributors or procuring relevant information. The team has been an ardent support to the editorial, designing and production team. Their endless efforts to recruit the best for this project, has resulted in the accomplishment of this book. They are a veteran in the field of academics and their pool of knowledge is as vast as their experience in printing. Their expertise and guidance has proved useful at every step. Their uncompromising quality standards have made this book an exceptional effort. Their encouragement from time to time has been an inspiration for everyone.

The publisher and the editorial board hope that this book will prove to be a valuable piece of knowledge for researchers, students, practitioners and scholars across the globe.

List of Contributors

Haji Karimi
Ilam University, Iran

V.J. Banks
British Geological Survey, Kingsley Dunham Centre, Nicker Hill, Keyworth, Nottingham, UK
University of Derby, Kedleston Road, Derby, UK

P.F. Jones
University of Derby, Kedleston Road, Derby, UK

Gholam A. Kazemi
Faculty of Earth Sciences, Shahrood University of Technology, Shahrood, Iran

Azam Mohammadi
Senior Hydrogeologist, Base Studies Section of the Water Resources Department, Northern Khorasan Regional Water Company, Bojnourd, Iran

Vincenzo Comegna, Antonio Coppola and Alessandro Comegna
Dept. For Agricultural and Forestry Systems Management, Hydraulics Division, University of Basilicata, Potenza, Italy

Angelo Basile
Institute for Mediterranean Agro-Foresty, (ISAFOM), National Research Centre, Ercolano, Italy

Milka M. Vidovic and Vojin B. Gordanic
The University of Belgrade, The Institute of Chemistry, Technology and Metallurgy, Department for Ecology and Technoeconomics, Serbia

Teresita Betancur V., Carlos Alberto Palacio T. and John Fernando Escobar M.
University of Antioquia, Colombia

Sultan Awad Sultan Araffa
National Research Institute of Astronomy and Geophysics, Helwan, Cairo, Egypt

Printed in the USA
CPSIA information can be obtained
at www.ICGtesting.com
JSHW011419221024
72173JS00004B/598